Fundamental Microbiology for Allied Health

Third Edition

H. Kathleen Dannelly, Ph.D.
Angela K. Chamberlain, M.S.
William M. Chamberlain, Ph.D.

Indiana State University

Printed in the United States of America.

ISBN: 1-60250-139-4
         978-1-60250-139-3

Table of Contents

To the Student

The laboratory exercises outlined in this manual are designed to give you hands-on experience in microbiology with special attention paid to the needs of the allied health professionals. You will learn to use a microscope to view bacteria, fungi and parasites. You will learn the proper techniques for culturing and studying these organisms, and in many of the exercises you will have the opportunity to practice your sterile technique.

We have emphasized the transmission of disease and included several exercises that demonstrate the ease with which microbes can be passed from one person to another. We have also included exercises that show effective means of controlling the spread of disease. Throughout the manual, emphasis is on the most common disease-causing organisms, where they are found, and how they are identified.

Safety is a priority in the microbiology laboratory; you will be presented with techniques that ensure the safety of everyone. It is imperative to learn and practice these techniques throughout the semester. On the first day of class the instructor will go over the rules and regulations for the class. Take care to follow these rules. They were developed for your safety and the safety of your classmates.

In many of the lab exercises you are required to draw what you view through the microscope. While this is not an art class, we do expect you to take special care to make realistic drawings. You will be required to provide colored pencils to complete the drawings.

Very often you will be expected to interpret your lab results. Remember, we are not just asking you to reiterate the results, but rather to tell what they mean. After reading the exercise you should have a very good idea what the interpretation should include, if not you should discuss this with your teaching assistant.

Questions have been placed at the end of each exercise to help in your studies. They were developed to help you understand the most important principles from that exercise. It would be prudent to answer and study these questions each week, and to do this work *yourself*. Each week your instructor will look over your answers from the last laboratory and will keep a record of whether you have answered the questions. This is to encourage you to keep up with the exercises. But even more importantly, you *need* to work with these questions to prepare yourself for quizzes and exams.

A *results page* accompanies each exercise. This is where you are expected to record your results. Periodically, the instructor will check to see that you have completed the results page at the end of the laboratory. We encourage you to do the work during the lab period while the information is fresh in your mind. When you go to work in an allied health field, you will need a good knowledge of sterile technique and you will be required to use these techniques on a regular basis. We believe you will find this to be an invaluable experience.

We have written this manual with the allied health student in mind. It is our best effort to provide exercises that will teach the causes of microbial diseases, routes of transmission, methods for controlling microbes, and the technical skills to work with bacteria and other important pathogens. We hope that you will find this manual gives you information and technical skills that you will carry with you throughout your career in allied health.

H. Kathleen Dannelly
Angela K. Chamberlain
Wm. M. Chamberlain
November 2006

ACKNOWLEDGEMENTS

We wish to thank our colleagues, teaching assistants and the students who have had input into the preparation of this lab manual. Their suggestions for improvement have been invaluable and their diligence in ferreting out typographical errors has saved us hours of tedious proof reading. Of the many who have participated special thanks is extended to Dr. Mary Ann McLean and Dr. Rosaline Waworuntu.

Dr. Maclean offered many useful suggestions regarding the content of certain labs and closely edited extensive portions of text.

As a teaching assistant Dr. Waworuntu provided invaluable feedback regarding student reactions to the lab exercises and her graceful hands are featured in the illustrations of proper laboratory technique.

In the final analysis, responsibilities for deficiencies in the lab manual rest with the authors and we accept responsibility for any shortcomings. However, we hope that despite deficiencies the manual helps allied health students develop an appreciation for the importance of microbiology.

Microbiology Lab Rules & Procedures

1. Place all personal belongings (coats, book bags, etc.) in the space provided as indicated by the lab instructor.

2. Laboratory coats are required and should be stored in the laboratory. **Do not take your lab coat out of the lab.**

3. Long hair must be tied back at all times while you are in the laboratory.

4. Decontaminate your work area with disinfectant BEFORE, AND AFTER, each lab session.

5. DO NOT eat, drink, chew gum, or apply make-up while in the lab. Food and beverages are not allowed in the lab.

6. DO NOT place paper, pencils, fingers, or other objects in your mouth.

7. DO NOT mouth pipette cultures or reagents.

8. Report any problems with the microscope to the lab instructor.

9. Immediately report injuries to the lab instructor.

10. If a culture is spilled, immediately notify the lab instructor.

11. Disposal of Materials:
    a. Place all contaminated materials (cultures, media, pipettes, used slides, and tubes) in specified containers.
    b. Never discard contaminated materials in regular trash containers.
    c. Never pour cultures down the sink.
    d. Petri plates, enterotubes, cotton swabs, pipettes, and other disposable materials should be placed in the biohazardous waste container.
    e. Used glass slides should be placed in the designated containers.

12. TURN OFF THE GAS BEFORE LEAVING THE LAB.

13. Return all reagents, wire loops, marking pens, microscopes, and other materials to their proper storage area before leaving the lab.

14. Before storing your microscope make sure it:
    ✓ Has the low power objective lens in the view position
    ✓ Is clean
    ✓ Has the cord properly secured
    ✓ Is returned to the appropriate storage cabinet

15. Wash hands thoroughly with soap and water before leaving the lab.

Section I: Microbiological Techniques

The study of microorganisms is dependent upon the use the proper techniques. Proper techniques include "tried and true" methods for slide preparation and staining, methods for isolating and growing microorganisms, and methods for identifying and determining their antibiotic susceptibility. Learning the proper techniques for handling cultures will allow the student the means to perform tests with pure cultures and thus obtain meaningful results.

A primary concern in the microbiology laboratory is the safety of laboratory personnel. Such safety is dependent upon the use of proper techniques for handling cultures. A single bacterial colony contains millions of cells and if mishandled, a colony of a pathogenic organism could infect many people. With proper techniques bacterial cells can be isolated on Petri dishes and transferred to other media without contaminating personnel or the environment. Thus it is very important that the techniques are learned well and practiced throughout the semester.

Another important consideration is that of uniformity in performing microbiological techniques. In a clinical microbiology laboratory it is especially important that experiments are performed in a standardized manner so that the results are comparable no matter who performs the test. There must also be uniformity between laboratories. In other words, it should not matter whether a throat culture is performed in Terre Haute, or New York, the results should be the same. To achieve these objectives, large volumes have been written on the standard procedures used in clinical laboratories throughout the United States. The same is true for other types of labs as for example those associated with water and sewage treatment facilities and with the EPA. Testing procedures follow standardized methods and government agencies monitor laboratories to ensure uniformity in the application of these techniques. As a result, potable water supplies in cities and towns throughout the U.S. are safe.

In this section, commonly used microbiological techniques are introduced and the theory behind the methods is discussed. While practicing these techniques the student and the instructor should pay close attention to the details of the procedures to prevent contamination of cultures, to ensure the success of the experiments, and the safety of the students.

Exercise 1: Introduction to Microscopy

Objectives: Throughout this course you will be using the compound light microscope to examine bacteria and other microorganisms. The following exercise is designed to explain how a microscope works and to teach the proper use and care of the microscope. Upon completion of this exercise you should be able to:

❖ Identify the major parts of the microscope and know the function of each.
❖ Know which objective lens to use when viewing eukaryotic specimens.
❖ Know which objective lens to use when viewing prokaryotic (bacteria) specimens.
❖ Define and calculate magnification.
❖ Define resolution.
❖ Know when and why immersion oil is used in microscopy.
❖ Know how to properly use, clean, and store the microscope.

Applications: In the clinical setting light microscopy is the most common method used for the immediate detection of microorganisms in patient specimens and for characterizing microbes grown in culture.

One of the most important and basic techniques required in microbiology is the effective use of the microscope. A typical microscope can magnify images approximately 1000X. This allows t observation of the size, shape, and motility of prokaryotic cells.

In light microscopy, light is transmitted through a specimen and then passes through a series of magnifying lenses. The compound microscope has two sets of lenses- the ocular and the objective. Your microscope has 4 objective lenses:

Objective Lenses of the Compound Light Microscope:
4X: scanning lens 10X: low power lens; useful for initial searching & focusing 40X: high dry lens; used to view most eukaryotic cells 100X: oil immersion lens; used to view bacterial specimens

The ocular lens has a magnification of 10X. The total magnification is equal to the objective magnification multiplied by the ocular magnification. Therefore:

Total Magnification:
Scanning: (4X)(10X) =   40X Low power: (10X)(10X) = 100X High Dry: (40X)(10X) = 400X Oil Immersion: (100X)(10X) =1000X

4

Two characteristics, the resolution and the magnification, determine the performance of a microscope. The usefulness of a microscope depends not so much on the degree of magnification but on the ability to clearly separate (resolve) two objects that are close together. Resolution is a measure of the clarity and detail of a magnified image and is measured as the shortest distance between two adjacent points that appear distinct, not a single blur. Magnification produces an image that appears larger and is measured as the ratio of image size, as seen through the microscope, to the actual size of the object. Magnification without detail is useless; two points that appear as a blur at low magnification will remain a blur at higher magnification. The best light microscope has a maximum resolution of 0.2 m and a practical magnification of 1000X- a resolving power approximately 500 times better than the resolving power of the human eye.

Proper illumination is essential for producing resolution. The function of the condenser is to focus and direct light from a source to the specimen. The position of the condenser, that is the distance between the condenser and the stage, determines how accurately the light is focused on the end of the objective lens. The iris diaphragm controls the amount of light entering the condenser. As the magnifying power of a lens increases, the diameter decreases and the lens admits less light. As a general rule, the amount of light being directed through the slide should be increased as the magnification is increased. Conversely, too much light may result in a loss of resolution and contrast.

To obtain maximum resolution when using high magnification, immersion oil is used to displace the air between the lens and the specimen. A small drop of oil is added directly to the stained smear and the end of the oil immersion lens is immersed in it. The oil serves to prevent the bending of light rays (refraction) that occurs when the light passes from glass (the slide) to air and back into glass (the lens). Refraction would reduce the amount of light entering the smaller opening of the oil immersion lens (100X), thus reducing resolution.

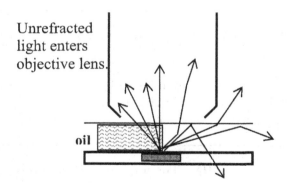

Figure 1.1 Light transmission with and without immersion oil

Some microscopes are designed to allow the operator to focus with low power and then adjust only the fine focus to obtain a clear image when another objective is moved to the viewing position. When a microscope has this capability it is called parfocal.

Routine Care of Microscope:

The microscope you will be using this semester is an expensive precision instrument that requires proper handling and care. Please keep in mind the following rules:

1. When carrying the microscope; always use two hands- one on the arm of the microscope and the other supporting the microscope under the base.
2. Keep the microscope flat on the lab bench and away from the edge.
3. Do not allow immersion oil on any part of the microscope other than the oil immersion (100X) objective lens. If, inadvertently, oil drips on the scope, remove it immediately with tissue paper and lens cleaner.
4. At all costs, do not get oil on the 40X objective. This is very important to remember. The 40X objective is almost the same length as the 100X, so if you are using oil immersion and you turn the nosepiece, moving the 40X objective into place, the tip of the objective will touch the oil. This destroys the 40X objective. They are not sealed to prevent oil from entering around the lens as are the oil immersion lenses.
5. Cleaning: Always use only the lens paper provided for cleaning the lenses to avoid scratches. You may need to moisten the lens paper with a few drops of lens cleaner especially when removing oil.
6. Storage: The microscope should be returned to the microscope cabinet when you are finished for the day.

   Do the following before putting away the microscope:

   -Use the coarse adjustment knob to move the stage away from the objective lenses.
   -Place the 4X objective lens in the view position.
   -Remove the slide from the stage & gently blot excess oil from the smear
   -Use lens paper to clean all lenses, cleaning the oil immersion lens last
   -Wipe any excess oil from the stage
   -Turn off power
   -Wrap the electrical cord on itself and hang it from the ocular.

6

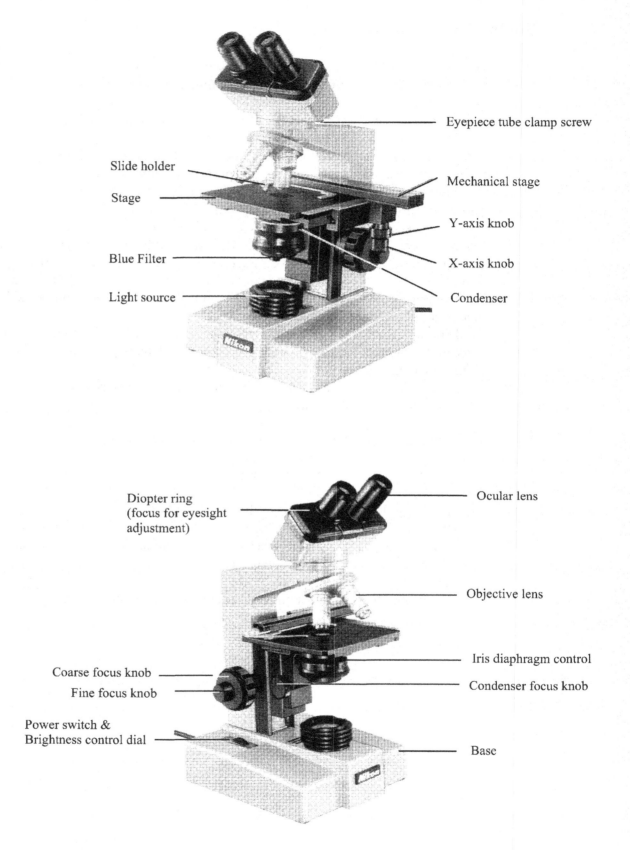

Figure 1.2 Parts of the microscope.

Table 1.1 Functions of the Parts of the Microscope

Ocular Lens	Eyepiece, magnifies image (10X)
Diopter Ring	Focusing ring located near bottom of left ocular for eyesight adjustment
Objective Lenses	Magnify the image (4, 10, 40, 100X)
Stage	Platform to support slide
Slide Holder	Secures slide on stage
Mechanical Stage	Allows movement of slide
X-axis Knob	Moves slide right & left across the stage
Y-axis Knob	Moves slide forward & backward across stage
Coarse Focus Knob	Rapidly brings specimen into focus
Fine Focus Knob	Slowly brings specimen into sharp focus
Condenser	Focuses light on specimen & fills lens with light
Condenser Focus Knob	Raises & lowers condenser
Iris Diaphragm	Controls amount of light leaving the condenser
Base	Supports microscope

Table 1.2 Brief guide for microscopy troubleshooting

Problem	Possible Cause(s)	Possible Solution(s)
Unable to see smear on slide well enough with the naked eye to center slide on stage.	Material on slide is too sparse.	Focus on landmark along the edge of the slide, then search for smear under low power.
Unable to focus on specimen under low power with the coarse adjustment.	Fine focus knob is turned too far.	Center the fine focus knob by turning it all the way in one direction and then back approximately midway. Then proceed by focusing with the coarse adjustment knob.
Wet mount specimen is foggy or barely visible.	Too much light on slide.	Adjust the brightness control dial. Adjust the iris diaphragm. Adjust the condenser focus knob.
Not enough light on specimen.	Brightness control dial is turned down, iris diaphragm is closed, &/or condenser is too far down.	Adjust brightness control dial, adjust iris diaphragm, adjust condenser focus knob.
Unable to obtain sharp focus using 100X objective lens.	No immersion oil. Not enough immersion oil. Condenser is too low.	Add oil. Adjust condenser.
Unable to find specimen on 100X though visible under 10X & 40X	Slide is upside down. Need to adjust fine focus.	Turn slide over and begin again. Readjust fine focus knob.
Cannot obtain sharp focus with high dry objective lens.	Oil or dirt is on outside of lens.	Clean the lens.
Field of view is not complete or round.	Objective lens is not clicked in place.	Click objective lens into proper position.
Darkness at periphery of viewing area or uneven brightness throughout the view.	Condenser is not adjusted properly.	Readjust condenser focus.

Procedure:
Parts & Function:

Identify the following parts of your microscope and note the function of each: (Refer to **Figure 1.2** and **Table 1.1**)

Ocular lens
Objective lens
Condenser
Iris diaphragm
Coarse Focus
Fine Focus
Mechanical Stage

Examination of Prepared Slide with Low Power and High Dry Objectives:

1. Obtain a microscope and one of the prepared eukaryotic slides.
2. Move the low power (10X) objective in place by twisting the revolving nosepiece.
3. Place the slide in the slide holder on the stage.
4. Move the mechanical stage so that the specimen is centered over the condenser
5. Adjust the condenser so that it is just below the stage.
6. Adjust the iris diaphragm so that it allows plenty of light into the condenser.
7. Look from the side of the microscope to ensure the lens does not touch the slide while moving the coarse adjustment until the stage is all the way to the top. The low power objective will be close to the stage but will not touch it.
8. Look through the oculars and slowly move the stage down. Stop when the specimen comes into view.
9. Adjust the spacing of the oculars until you see one image.
10. Close your left eye and use the fine focus adjustment knob to focus for the right eye. Close your right eye and use the focusing ring located near the bottom of the left ocular to focus for your left eye. Open both eyes and adjust the fine focus adjustment until the image is sharp.
11. Open and close the iris diaphragm, lower and raise the condenser, and note what effect these adjustments have on the appearance of the image. Usually the condenser should be kept at the uppermost position.
12. Rotate the nosepiece so that the 40X objective is in place. Move the fine focus adjustment until the image is sharp and then examine your specimen.
13. Draw and describe what you observe.
14. When finished observing the slide, move the 4X objective into place, lower the stage and switch slides.
15. Look at 2 different eukaryotic demonstration slides. **Practice until you are comfortable using the microscope and it is easy to perform the routine steps in adjusting the scope to view a specimen.** Do not be alarmed if it takes several days of use and several hours of practice before you feel comfortable with finding specimens under the scope. Just remember, we expect you to work until you become comfortable.

Wet Mount Preparation—Compost Infusion

1. Place a drop of compost or hay infusion on a clean glass slide.
2. Place the edge of a coverslip at a 45° angle along one edge of the water droplet
3. Gently lower the coverslip over the drop of water. Avoid trapping air bubbles.
4. Observe under the scanning lens followed by high dry (40X)
5. You may need to reduce the amount of light by lowering the condenser to improve contrast.
6. Draw and describe what you see.

Oil Immersion—Examining Bacterial Slides

1. Obtain a bacterial slide.
2. Follow the procedure outlined for viewing slides under high dry to locate the smear.
3. Rotate the nosepiece so that none of the objective lenses are in position for viewing and the low power is closest to the mechanical stage mechanism. (NOTE: If the microscope is configured as is shown in **Figure 1.2** the 40X objective will be to the right of the view position and the oil immersion lens is to the left.).
4. Place 1-2 drops of immersion oil on the specimen.
5. Rotate the nosepiece so that the oil immersion lens (100X) is in place. Adjust (looking from the side) the objective position so that just the tip is touching the oil.
6. Avoid getting oil on anything other than the slide and the oil immersion objective lens.
7. The condenser should be all the way up to the stage when using the oil immersion lens.
8. The iris diaphragm should be all the way open.
9. Adjust the fine focus until you have a sharp image.
10. Draw and describe what you observe.

Results

Draw and label your observations using the high dry and oil immersion objectives.

High Dry: Name the organisms you viewed and give the total magnification for each drawing.

_____   _____

Total Magnification: _____   **Total Magnification:** _____

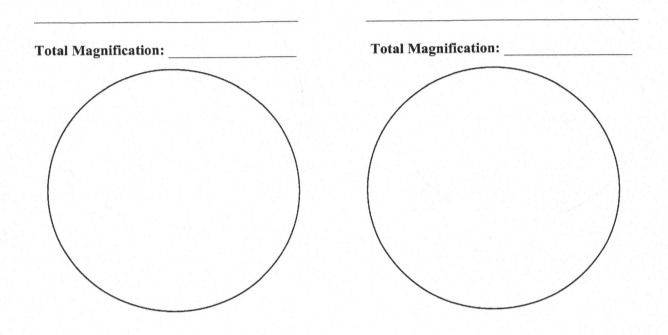

Wet Mount: Make drawings of two views of the material seen in the wet mount.

Total Magnification: _____   **Total Magnification:** _____

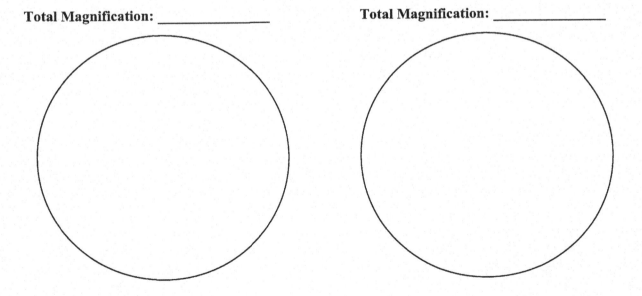

Oil Immersion: Make two drawings of the bacteria seen under oil immersion.

Total Magnification: _____   **Total Magnification:** _____

Questions

1. Which objective lens would you use to observe bacteria? Eukaryotic cells?

2. If the ocular lens on your microscope had a magnification of 5X, what would be the total magnification of the microscope with each of the objective lenses?

3. What two characteristics determine the performance of a microscope?

4. Define resolution.

5. What is the purpose of immersion oil?

Exercise 2: Microorganisms in Our Environment

Objectives: Observe the ubiquity and diversity of microorganisms in the environment. Upon completion of this exercise you should be able to:

- ❖ Define ubiquity.
- ❖ Explain why it is important to be aware of microbial ubiquity in the laboratory.
- ❖ Properly label Petri dishes.
- ❖ Explain why we incubate Petri dishes upside down.

Applications: Microorganisms are widely distributed in the environment. While the ubiquitous nature of microorganisms is not of great concern to healthy humans who practice good hygiene, it is a serious issue in the healthcare setting where susceptible patients must be protected from nosocomial (hospital-acquired) infections. As well, it is a serious consideration to laboratory personnel who must use sterile technique to maintain the sterility of media and supplies while culturing and identifying microorganisms.

UBIQUITY
Microorganisms are ubiquitous, meaning they are found virtually everywhere. Soil and water are teeming with microbes; microorganisms can adhere to dust particles, become airborne, and spread for miles; and the human body is virtually covered with microorganisms that naturally grow on skin and mucous membranes. It has been estimated that at least 99% of the microbes on earth are harmless to humans and, in fact, several types perform very important functions in the environment such as decomposition of dead plants and animals. One group of organisms, the nitrogen fixers, transforms nitrogen from the atmosphere into a usable form for plants and ultimately for animals. Without these organisms, life on earth could not exist.

The base of most microbial media is a nutritious broth containing the same nutrients as the foods we eat. It is sterilized with heat before use but to keep it sterile, it must be protected from contamination by not touching it or leaving it open to the air. Understanding that microbes are virtually everywhere, on the tabletop, in the air, on your skin, hair and clothing, can help your laboratory technique.

DIVERSITY
Microorganisms are extremely adaptable and this results in great diversity. Studies have shown that through mutation as well as acquisition of new genes from other organisms, microorganisms can change and thrive under almost any circumstance. Consequently microorganisms have been detected even in the most extreme environments on earth.

While the pH of most environments ranges from 6 to 8, natural areas with either a very low or very high pH do exist. In acid lakes and bogs a pH as low as 3.5 is typical while acid mine drainage resulting from mining disturbance can have a pH as low as 1.0, a pH as low as the pH found in the stomach of mammals. In all of these environments, there are microbes that are especially adapted to thrive.

On earth we find extremes in temperature as well. The heat of the tropics or the freezing cold of the Antarctic are obvious. But even in hot springs, geysers, and the thermal vents in the oceans, where temperatures verge on boiling, microbes live and grow. One of the trade-offs for an organism adapted to the extreme is loss of the ability to live in non-extreme areas. The term extremophile was coined to describe organisms that have become dependent upon extreme conditions for life.

In addition to the extremes, microorganisms also adapt to even subtle differences in their chosen environments. With many of the microbes that live at neutral pH, even a change of 0.5 pH units from their optimum can slow there growth rate to half its potential. A small change in conditions may allow other organisms to out-compete and thus dominate the environment

The temperature optimum of an organism (the temperature at which it grows best) often dictates where it can survive Many microbes are adapted to grow best at what laboratory scientists call "room temperature" that is 20 to 22°C (68-72°F). It is not surprising then that many of the human pathogens grow best at body temperature (37°C or 98.6°F). *Neisseria gonorrhea* for example is very particular and must have a temperature of 35-37°C to survive it is highly adapted to growth in the human body and is not found anywhere else. On the other hand, *E.coli* (found in the intestine of most animals including humans) can grow in a wide range of temperatures, from 15-37°C, but grows fastest at 37°C. And then there are some, such as *Mycobacterium leprae* (the cause of leprosy), that are restricted to growth in another organism because of their nutritional requirements but grow best at a lower temperature: consequently, *M. leprae* is found growing in the cooler parts of the skin, the face, buttocks, and extremities.

The temperature at which organisms are isolated varies depending upon their optimum growth temperature. For example: if a skin culture is incubated at room temperature and observed after 24 hours, there may be no growth on the plates. If the same sample is plated and incubated at body temperature (37°C), after 24 hours many organisms will have grown.

Growth media is also important in growing microorganisms. Certain media favor the growth of certain organisms. For example, in this exercise you will be using malt extract agar (MEA) which has a lower pH (4.5-5.0) than typical nutrient agar (pH ~7). Most fungi are able to tolerate the lower pH and are therefore more likely to grow on this type of media than are bacteria. It is important to understand your samples prior to plating and incubation so the conditions supplied are most favorable for the organisms you wish to isolate.

Procedure:

1. Always label each Petri dish on the bottom (agar side) before you begin an experiment. The lids of the plates are identical and can be easily mismatched.

2. During the laboratory session the Teaching Assistant (TA) will assign each pair of students one of the following exercises to demonstrate microbial ubiquity:

   a) Place an open agar plate in the 1) restroom, 2) hallway, or 3) on your lab bench for 30 minutes.
   b) Place an open agar plate on your lab bench for the whole lab period.
   c) Touch the surface of the agar with your fingers.
   d) Cough on the surface of an agar plate.
   e) Rub the surface of your hair with a sterile swab and streak the surface of the agar.
   f) Swab an area of your lab bench and streak the surface of the agar.
   g) Swab an area of the floor and streak the surface of the agar.
   h) Swab the sink/ faucet and streak the surface of the agar.
   i) Swab the handle of a soap dispenser and streak the surface of the agar.
   j) Swab the surface of a coin or paper bill and streak the surface of the agar.

3. To transfer a sample from a cotton swab to a agar plate, gently move the cotton swab over the surface of the media as shown in **Fig. 2.1**. Take care not to push too hard and break the surface of the agar. Occasionally twist the swab between your fingers to ensure that all sides of the swab touch the agar.

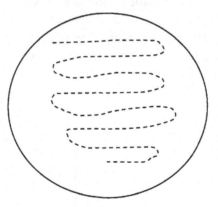

Figure 2.1

4. Incubate plates at room temperature in the lab. Turn plates upside down to incubate. This is done so that condensate which accumulates on the lid of the Petri dish does not drip onto the surface of the agar and increase the chance that the plate becomes contaminated. Water on the surface of the plate also allows motile organisms to easily travel across the agar surface preventing the isolation of pure colonies.

5. You will be given the following materials to take home:

> 1 malt extract plate
> 2 nutrient agar plates
> parafilm strips
> sterile swabs

6. Use the items listed above to obtain environmental samples. Be creative in your sampling and take four samples within the next 24 hours. Divide your nutrient agar plates into four quadrants by drawing a cross on the bottom of the plate as shown in **Figure 2.2**. By swabbing the surface of the agar, inoculate each quadrant of a nutrient agar plate with a sample from a different environment. Then using the same set of swabs, inoculate in a similar fashion, the second nutrient agar plate. You should end up with both nutrient agar plates inoculated identically with the same four samples. For each quadrant record the source of the sample and be careful not to mix the samples. Once you have completed your sampling, secure each plate with Parafilm by stretching and fitting the Parafilm around the edge of the plate. Store your plates in the refrigerator until the day of your lab. Bring the plates to your next lab session. One nutrient agar plate will be incubated at room temperature and the other at 37°C.

7. The malt extract plate may be used in the same way as the nutrient agar plates, swabbed or used to sample the air in your home or dorm room. Damp basements or outdoors (gardens, lawns, compost heaps, forests) may be particularly rich in fungi. Sample by swabbing as described above or, for air samples, open the plate for 30 min in the environment of interest. Close the dish and seal with Parafilm. Incubate the malt extract plate at room temperature for 3-5 days according to the lab instructor's directions.

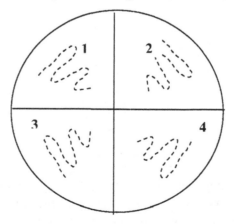

Figure 2.2

Results

1. Compare the growth on the nutrient agar plate incubated at 37°C with that on the plate incubated at room temperature (22°C). Did you get what you expected? Did the swabs of human surfaces produce more colonies at 37°C or at room temperature? Did the swabs from the environment (doorknobs, household items etc) produce more colonies at 37°C or 22°C? Record your results.

2. If possible, count the number of colonies on each quadrant of each plate. Record the results.

3. If you assume any observable difference in colony characteristic (shape, color, size) indicates a different species, how many different species grew on your plates?

4. Is there a predominance of one or two types of colonies on any of your plates? If so, describe.

5. The malt extract plate contains nutrients that promote fungal growth. Compare your results with the results obtained by other students who followed similar sampling procedures. If air was sampled, which environments resulted in higher numbers of colonies per plate? If surfaces were swabbed, which surfaces produced more fungal colonies per plate? Speculate as to why this might be.

Questions

1. Where should a Petri dish be labeled and why?

2. Why is it important to be aware of the ubiquity of microorganisms while working in the laboratory?

3. What is the reason for incubating plates upside down?

Exercise 3: Bacterial Colony Morphology

Objective: Learn how to recognize and describe the variations in bacterial colony morphology. Upon completion of this exercise you should be able to:

- ❖ Define colony.
- ❖ List the characteristics used to describe bacterial colonies.
- ❖ Describe an unknown bacterial colony.
- ❖ Explain why the production of isolated colonies is an important first step when identifying bacteria.

Application: Colony appearance is often distinctive for particular bacterial species and is useful in the initial stages of identification.

A colony is a visible mass consisting of millions of bacterial cells. Each individual colony arises through multiple divisions beginning with a single cell; each colony represents the progeny of a single cell and is therefore considered a pure culture. The appearance of the colonies (size, shape, color) is dependent upon the species or strain of bacteria and substances they produce. Therefore, colony morphology is genetically determined and is often the first step in the differentiation and identification of bacterial species. Important observable characteristics that are useful in describing colonies are:

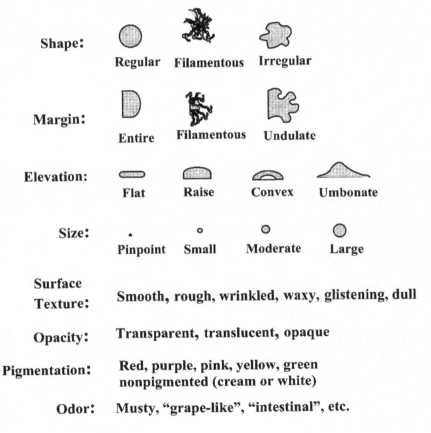

Shape: Regular Filamentous Irregular

Margin: Entire Filamentous Undulate

Elevation: Flat Raise Convex Umbonate

Size: Pinpoint Small Moderate Large

Surface Texture: Smooth, rough, wrinkled, waxy, glistening, dull

Opacity: Transparent, translucent, opaque

Pigmentation: Red, purple, pink, yellow, green nonpigmented (cream or white)

Odor: Musty, "grape-like", "intestinal", etc.

Figure 3.1 Bacterial colony characteristics

Procedure:

1. Observe the demonstration plates provided. Note the various colony types. Using the terminology given in **Figure 3.1** write an appropriate description for 3 of the colony types found on the demonstration plates. Record this information in **Table 3.1**, the **Results** section.

2. Using the terminology given in **Figure 3.1** write an appropriate description for 3 of the colony types growing on the plates you obtained by sampling the environment. Record your descriptions in **Table 3.2**, the **Results** section.

Results

Table 3.1 In the table below give the name and colony morphology of 3 demonstration organisms.

	Known Organisms		
Name			
Shape			
Margin			
Elevation			
Size			
Texture			
Opacity			
Pigmentation			

Table 3.2 In the table below describe the colony morphology of three unidentified organisms obtained from the environment.

Isolate	Unknown #1	Unknown #2	Unknown #3
Shape			
Margin			
Elevation			
Size			
Texture			
Opacity			
Pigmentation			

Questions

1. Define the term colony as used in microbiology.

2. Why is it important to observe and describe the morphology of a colony?

3. Why would it be important to first isolate a colony when trying to identify an unknown bacterial culture?

Exercise 4: Aseptic Techniques, Isolation & Transfer of Pure Cultures

Objective: Learn aseptic techniques necessary for transferring bacterial cultures. Learn how to obtain isolated colonies via the streak plate technique. Upon completion of this exercise you should be able to:

❖ Define pure culture.
❖ Give examples of two situations in which pure cultures are necessary.
❖ Define aseptic technique.
❖ Define culture medium.
❖ Aseptically transfer bacteria from an agar plate culture to a broth medium.
❖ Aseptically transfer bacteria from an agar plate culture to an agar slant.
❖ Aseptically transfer bacteria from a broth culture to agar medium.
❖ Explain how a correctly streaked plate produces pure cultures.
❖ Successfully obtain isolated colonies by performing a streak plate.
❖ Explain the importance of proper aseptic technique in the laboratory.

Application: In most areas of microbiology, the ability to obtain pure cultures is critical. In the clinical setting obtaining a pure culture helps determine the various types of bacteria present, aids in identifying the organisms most likely to cause disease, and facilitates tests to determine the effects of antibiotics used for treatment.

Microscopic observation gives us important details regarding the appearance of bacteria (cell shape, size, and arrangement). By studying the growth of bacteria in laboratory culture media we can learn much about their biological activities.

Samples from the environment and patient specimens most often contain mixed populations comprised of several bacterial species. Before important characteristics of a particular bacterial species can be determined, the organism must be isolated in pure culture. A pure culture is a culture consisting of only one type of microorganism, a population consisting of individuals derived from a single organism.

To develop and maintain pure cultures, techniques for transferring growing organisms from a pure culture to a sterile medium without introducing contaminants are essential. Methods for manipulating cultured organisms while preventing contamination are called aseptic techniques.

A sterile solution containing the necessary nutrients for bacterial growth is called a culture medium (plural=media). Media can be prepared as either liquids (broths) or solids (agar slant, agar deep, or agar plate). Solid medium is produced by adding agar, a gelatinous material extracted from seaweed, to hot nutrient solution. When cool, the agar gels to form a soft solid, similar in consistency to gelatin desserts.

A. Streak Plate Technique

Even a small sample from a bacterial colony contains thousands of cells. To obtain a pure colony, such a sample must be diluted. A simple and rapid method of obtaining pure cultures involves the mechanical dilution of a small sample on an agar plate. This method is called the **streak plate technique**. To apply this technique an inoculating loop containing a bacterial sample is moved gently, yet rapidly with a flinging motion, across the surface of an agar plate following a particular pattern (see **Figure 4.1**). As the loop moves across the agar, bacteria are deposited and the cell density decreases. Eventually individual cells become separated and deposited singly on the agar surface. After the plate has been incubated, the area at the beginning of the streak pattern will show confluent or near confluent growth of bacteria. Confluent growth is when there are no distinct colonies; rather there are so many colonies growing so close together that they run into each other and form solid growth. In the area near the end of the pattern you should find discrete colonies. These discrete colonies are pure colonies consisting of a large group of cells derived from a single individual.

Before you begin using your wire loop for any activity, you will need to sterilize it. This is done by holding the wire in the flame of a Bunsen burner until it becomes red hot. It is important to give the loop approximately 30 sec to cool before touching a bacterial colony or you will kill the bacteria. The reason for flaming the loop between quadrants of a streak plate is to aid in the dilution of the bacteria by removing all of the residual bacteria on the loop. You must then be sure to pull bacteria from the last quadrant streaked into the new quadrant. **Figure 6.3** shows the proper way to hold the wire in the flame to allow as much of it to become red hot as possible.

B. Transfer of Pure Cultures

There are many reasons for transferring pure cultures in the microbiology lab among which are maintaining stock cultures and transferring to identification media. Good sterile techniques are essential for the transfer process to maintain a pure culture. And the transfer techniques must include procedures for transferring to, and from, solid and liquid media.

Solid to Solid Transfers (see Figure 4.3)
In this exercise you will transfer organisms from a pure culture grown on a plate to an agar slant in a test tube. The tubed agar media is slanted prior to solidification to give more surface area for inoculation. When culturing from solid media t is not essential to obtain a visible amount of colony on your loop. If you merely touch a colony with your loop or swab, it will pick up a sufficient number of organisms to obtain growth in the new media. When transferring to tubed or plated media, it is important to imagine it is raining bacteria in the lab. Keep the media surface as dry as possible and only lift the lid or remove the cap for as long as is absolutely necessary to make the inoculation.

Using a loop or swab, touch a pure colony to obtain an inoculum for your new culture. It is not necessary to streak the new media for isolation since you are starting with a pure culture. Remove the cap, touch the surface of the agar with the loop or swab, and make a squiggle line up the surface of the slanted agar. The tubed agar media is slanted prior to solidification to give more surface area for inoculation. Replace the cap and incubate.

Liquid to Solid Transfers (see Figure 4.3)

When transferring liquid cultures, you may use either a sterile loop or swab. Surface tension of the water/media will allow you to pick up a droplet in the looped wire and carefully remove it from the source. The droplet can be transferred to either a liquid or to a solid media. When transferring to a slant, a squiggle line is made on the surface of the slant with the loop or swab. When transferring to a plated media, the droplet is applied to one quadrant of the plate. The plate is then streaked to obtain isolated colonies or confluent growth.

Figure 4.1 Streak Plate Method of Isolation

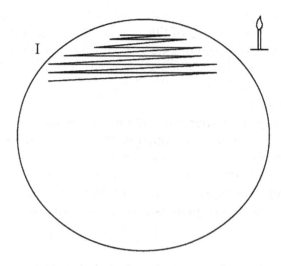

1. Streak the culture in one quadrant of the agar plate. Flame the loop.

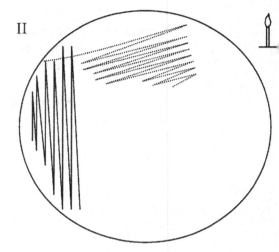

2. Rotate the plate~90°. Cool the loop by touching it to an uninoculated region of the agar. Streak quadrant II. Flame the loop.

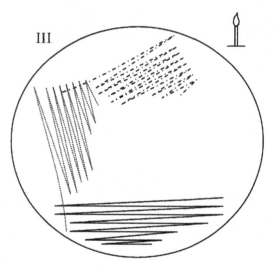

3. Rotate the plate~90°. Cool the loop by touching it to an uninoculated region of the agar. Streak quadrant III. Flame the loop.

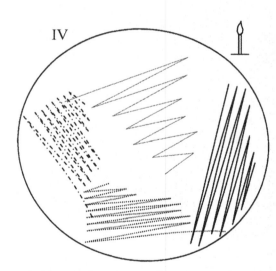

4. Rotate the plate~90°. Cool the loop by touching it to an uninoculated region of the agar. Streak quadrant IV. Flame the loop. Incubate the plate.

Figure 4.2 Illustration of the growth pattern on a typical streak plate.

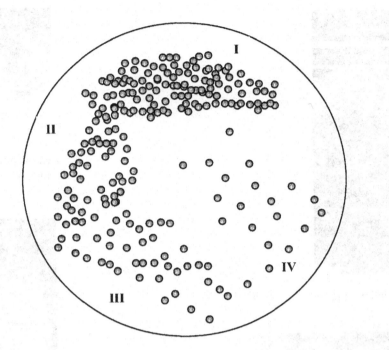

Figure 4.3 Culture transfer techniques

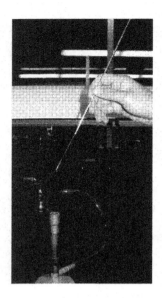

(a) Flaming inoculating loop

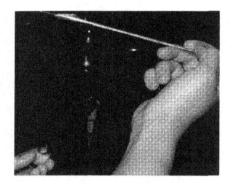

(b) Proper removal of culture tube cap

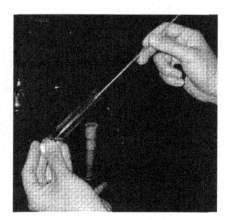

(c) Removal of liquid culture using inoculating loop

(d) Removing culture from plated agar media

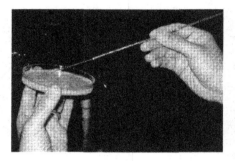

(e) Inoculation of plated agar media

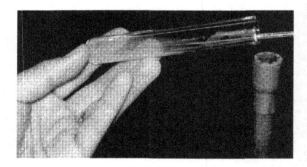

(f) Inoculation of tubed agar media

Procedure:

1. Refer to the procedural diagrams in **Figures 4.1**, **4.2** and **4.3** and use proper aseptic technique to perform the following culture transfers:

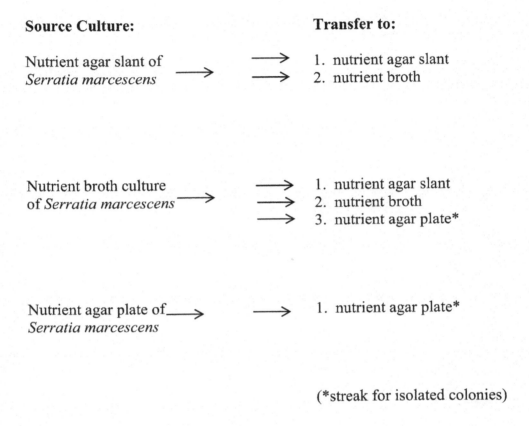

Source Culture: **Transfer to:**

Nutrient agar slant of   1. nutrient agar slant
Serratia marcescens   2. nutrient broth

Nutrient broth culture   1. nutrient agar slant
of *Serratia marcescens*   2. nutrient broth
                                          3. nutrient agar plate*

Nutrient agar plate of   1. nutrient agar plate*
Serratia marcescens

(*streak for isolated colonies)

Results

1. Describe the growth that appears on the various media types. (Do all the colonies look alike? Are there isolated colonies? etc.)

    a. Slants –

    b. Broths –

    c. Agar plates streaked for isolation –

Questions

1. Define pure culture.

2. Give 2 examples of situations in which it is important to obtain a pure bacterial culture? Explain why.

3. Why is proper aseptic technique so important in the laboratory setting?

Practical Exercise 4A: Streak Plate Method

For this practical test you will be given a Petri plate with nutrient agar, a culture of *E. coli* on a nutrient agar slant, a marker, a loop, and a Bunsen burner. The test consists of performing a streak plate from beginning to end while your teaching assistant observes your technique. When the teaching assistant instructs you to begin, you are expected to use aseptic technique to transfer the organism to the Petri plate and then streak for isolation. Half of your grade will be based on your demonstration of sterile techniques, proper use of the loop, etc. and half your grade will be derived from the results of your streak plate. (The considerations will be 'Did you perform the streak correctly?' and 'Did you get isolated colonies?').

42

Practical Exercise 4B: Aseptic Techniques

The term aseptic literally means without infection. Aseptic techniques are methods used to prevent contamination by microorganisms. Medical personnel use aseptic techniques in dealing with patients to prevent contamination of surgical sites, wounds, catheters, etc. As you perform this exercise you should keep in mind previous exercises that demonstrated the fact that microorganisms are everywhere including in the air as dust and other aerosols.

Procedure:

1. You will be given two pairs of sterile surgical gloves.

2. One pair will be used for sterility testing following a procedure used in the latex glove industry to test their product for sterility.
    a. Carefully open a pair of sterile latex gloves.
    b. Remove forceps and scissors from the ethanol and burn off the ethanol taking caution not to get the instruments hot*.
    c. With the forceps, pick up the tip of one of the fingers of the gloves and cut off a ~0.5 inch long piece.
    d. Using aseptic technique, place the piece you removed in the broth tube provided.
    e. Incubate overnight at 37°C and observe for growth of any microorganisms.

3. The second pair will be used to determine if you are able to put on the gloves without contamination. Your TA will demonstrate the proper aseptic technique.
    a. Open the second pair of gloves. Pick up one glove and put it on, touching it only on the folded back (inside) cuff.
    b. Once the first glove is on use your gloved hand to pick up the second glove, touching it only on the outside; this is done by placing the gloved fingers under the folded cuff.
    c. To maintain sterility of your gloves, have your lab partner open the Petri plate when you have both gloves on. To demonstrate that the gloves are still sterile touch the surface of the agar.
    d. Incubate the plate at 37°C and observe for growth of any microorganisms.

* 95% ethyl alcohol can be used to sterilize objects such as scissors and forceps. The difficulty is that the ethanol will inhibit the growth of microorganisms if it is allowed to contact the substance being tested or the growth media. This difficulty is overcome by lighting the ethanol with a Bunsen burner and allowing the ethanol to burn away. DO NOT HOLD THE INSTRUMENTS IN THE FLAME. Ignite the alcohol and pull the instrument out of the flame to burn. It is not necessary or desirable that the instruments become hot. Remember they are sterilized by the ethanol, not the flame as your loop is sterilized.

Exercise 5: Bacterial Smear Preparation & Simple Staining

Objectives: Learn how to prepare bacterial smears for staining purposes. Understand the basic principles of simple staining. Learn to recognize and describe bacterial cell shapes. Upon completion of this exercise you should be able to:

- ❖ Describe how to prepare a bacterial smear.
- ❖ Explain why we heat fix (flame) a bacterial smear.
- ❖ Explain why a thick smear is undesirable.
- ❖ Explain why we stain bacteria.
- ❖ Recognize and describe bacterial cell morphologies.

Applications: Information learned from staining and observing bacteria is useful in identifying bacteria. Proper smear preparation is often the key to successful staining of microbes.

Simple Staining:
Because bacteria are very small and almost completely transparent they are very difficult to observe without staining. The use of a single stain to color bacterial cells is called simple staining. The outer surface of most bacteria is negatively charged, so positively charged molecules are attracted to, and bind with, the bacterial cell surface. In general, basic (positively charged) dyes are used as direct simple stains. Stains color the bacterial cells to create contrast between the bacteria and the background so that the cells are clearly visible with the microscope. With simple staining, cell shape, size, and arrangement can be observed. Stained bacteria can be measured for size and are classified by their shapes and groupings.

Bacterial Smear Preparation:
The first step in the staining process is to prepare a smear; a dried preparation of bacterial cells on a glass slide is called a smear. The purpose of making a smear is to adhere/fix the bacteria to the slide to prevent the sample from being lost during the staining process. Smears can be prepared from liquid cultures or from cultures grown on an agar medium.

A. When using a liquid culture, one loopful of culture is smeared onto a glass slide and allowed to air dry. The cells in the dried smear are then "fixed" to the slide by briefly heating. This process is known as heat fixation.

B. When using growth from an agar medium, a loopful of water is placed on the slide and a very small amount of culture is mixed with the water to separate and suspend the bacteria. The suspension is then spread out, air dried, and heat fixed. In a good smear, the bacteria are evenly spread out on the slide and individual organisms are visible microscopically.

Bacterial Cell Morphology:

Bacteria have rigid cell walls that function to maintain a constant shape. There are three basic shapes; cocci, bacilli, and spiral. Bacterial cells group together as they multiply and the arrangement of these groups is often characteristic of a genus or species. See **Figure 5.1**.

Figure 5.1 Common shapes and arrangements of bacterial cells.

Coccus	**Bacillus**	**Spiral**
diplococcus tetrad streptococcus staphylococcus	diplobacillus streptobacillus	spiral "corkscrew" vibrio "coma-shaped"

(a) cocci

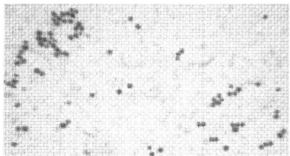

(b) bacilli

(c) spiral

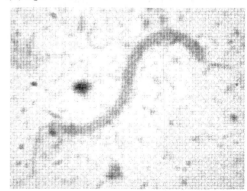

47

Figure 5.2 Bacterial smear preparation

1. Place a small drop of water on a clean microscope slide*.

2. With the inoculating loop aseptically add bacterial culture to the drop of water.

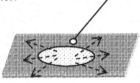

3. Mix the bacteria with the water to a fine suspension and spread it out.

4. Allow the suspension to air dry.

5. Heat fix the suspension (smear) by quickly passing the slide through a flame 2 times. Note: Slide should not be too hot to touch.

Figure 5.3 Procedure for simple staining of bacterial smears.

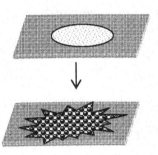

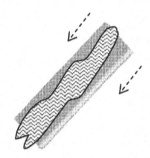

1. Cover prepared smear with stain.

2. Let sit 1 minute.

3. Rinse slide thoroughly with water.

4. Gently blot dry with paper towels. Do not rub.

5. Observe stain under oil immersion.

* Some students find it helpful to draw a circle with a wax pencil on the underside of the slide to mark the place where they are going to make the smear.

Procedure:

1. Choose an isolated/pure colony on one of the plates you obtained by sampling the environment.

2. Follow the procedure diagrammed in **Figure 5.2** and prepare a smear using the isolated colony as the source.

3. Follow the procedure diagrammed in **Figure 5.3** and stain the smear with methylene blue. Expose the smear to stain for approximately one minute.

4. Use the oil immersion lens to observe the material on the slide.

5. Use *Bacillus cereus* as the test organism and repeat steps 1-4 above.

6. Record your observations on the **Results** page.

Results

For each of the test organisms answer the following questions:

1. For each organism, describe what your stained smear looks like from the macroscopic view (to the unaided eye). Is it thick and dark blue? Or is it thinned out and a more faint blue?

2. What do you see when you first view your smear with low power? Describe it.

3. Describe what you see when you view your smear on high dry?

4. Find a field of view where the cells appear to be in a single layer. Draw what you see when you view your slide through the oil immersion lens.

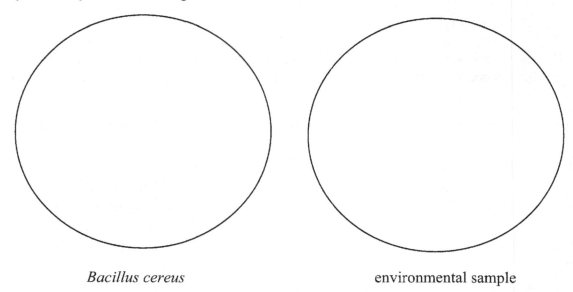

Bacillus cereus   environmental sample

5. Are there areas on the slide where the cells are heaped up into multiple layers? (This is the most common mistake made by beginning microbiology students.) You may notice that in these areas the cells did not stain properly and there are very large dark blue areas where no cells can be discerned under the scope. The next stain you prepare take care not to get the smear too thick.

Questions

1. What is the purpose of staining bacterial cells?

2. What is the purpose of heat fixing a bacterial smear?

3. What are the characteristics of a good bacterial smear?

52

Exercise 6: Gram Stain

Objective: Learn the Gram staining procedure and be able to distinguish between Gram-positive and Gram-negative bacteria. Upon completion of this exercise you should be able to:

❖ Define and compare simple and differential stains.
❖ Describe the difference(s) between Gram-positive and Gram-negative cell walls.
❖ Describe how the Gram stain is used to differentiate between Gram-positive and Gram-negative bacteria.
❖ Describe each step of the Gram stain procedure and indicate what color Gram-positive bacteria and Gram-negative bacteria are at the end of each step.

Application: The Gram stain is the most important and widely used stain in microbiology. It is the first differential test run on a specimen brought into the clinical laboratory for identification.

Bacterial species differ from one another chemically and physically and may react differently to a given staining procedure. Differential stains allow you to detect differences between organisms or differences between parts of the same organism. Differential stains are used much more frequently than simple stains because they allow the determination of cell size, shape, and arrangement, just as simple stains do, but also provide information about other cell characteristics. The Gram Stain is a differential stain.

In 1884, Christian Gram devised a technique that divided bacteria into two groups on the basis of their staining properties. These two groups are referred to as the **Gram-positive** and **Gram-negative** bacteria. Although it was many years before the mechanism was understood, the structure of the cell wall is different in these two groups and accounts for the different staining characteristics. A Gram-positive bacterium has a thick cell wall that consists primarily of a protein-sugar complex called peptidoglycan. A Gram-negative bacterium has a much thinner cell wall, also made of peptidoglycan, which is surrounded by an outer membrane with high lipid content. See **Figure 6.1**

The steps involved in the Gram stain procedure are outlined in **Figure 6.2**. Understanding the mechanisms involved in each step should help you perform the stain correctly. Crystal violet (the primary stain) is first applied to the smear. The cell walls of both Gram-positive and Gram-negative organisms stain purple with crystal violet. An iodine solution is next added to the smear. The iodine acts as a mordant and causes the crystal violet to form large crystals within the cell wall. The decolorizer (alcohol/acetone solution) dissolves the crystal violet and at the same time dehydrates the thick cell wall of the Gram-positive cells. This dehydration makes it difficult for the stain to leave the cell wall so the Gram positive cells remain purple. The effect of the decolorizer on Gram-negative cells is quite different. The alcohol dissolves the lipid in the outer membrane and washes the crystal violet out of the thin peptidoglycan layer. Therefore, the Gram-negative bacteria become colorless. A gentle wash with water ends the decolorization reaction. The timing of the decolorization step is extremely critical. If the decolorizer acts for too long a time, the crystal violet will wash out of the Gram positive cell wall and they too will become colorless. The addition of safranin (the counterstain) stains all cells pink but in the case of Gram positive cells, the pink color is obscured by the strong purple of the crystal violet remaining in the cell wall.

The Gram stain is the most widely used staining procedure in microbiology. It is typically the first in a series of tests performed to identify a bacterial species.

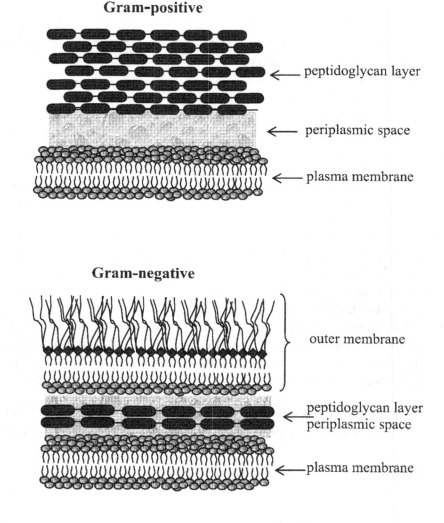

Figure 6.1 Gram-positive and Gram-negative cell wall components.

Figure 6.2 Gram Stain Procedure

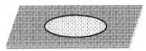

1. Start with a bacterial smear. See **Figure 5.2**.

2. Cover the smear with crystal violet stain for 1 minute.

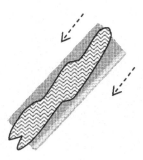

3. Gently and thoroughly rinse the slide with water.

4. Cover the smear with Gram's iodine for 1 minute.

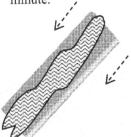

5. Gently and thoroughly rinse the slide with water.

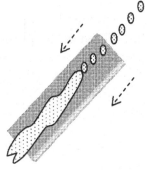

6. Add decolorizer drop wise until runoff is clear (10-20 seconds. Do not over decolorize.

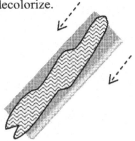

7. Gently and thoroughly rinse the slide with water.

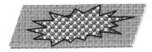

8. Cover the smear with safranin stain for 1 minute.

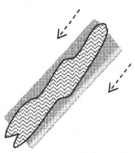

9. Gently & thoroughly rinse the slide with water.

10. Gently blot the slide dry with paper towels.

11. Observe stain under oil immersion.

Procedure:

1. Refer to the "Procedural Diagram" in **Figure 6.2** and perform a Gram stain on the following samples:

   -*Staphylococcus aureus*
   -*Escherichia coli*
   -Unknown sample

2. Observe the Gram stained slides under oil immersion.

3. Draw your observations on the **Results** page.

.

Results

Draw the results of your Gram stains below and fill in the blanks for each.

Escherichia coli

Gram ___

Shape _____

Arrangement _____

Staphylococcus aureus

Gram ___

Shape _____

Arrangement _____

Unknown

Gram ___

Shape _____

Arrangement _____

Questions

1. In what way(s) are differential stains more "useful" than simple stains?

2. Describe in detail what physical properties account for the different staining characteristics of Gram-positive bacterial cells vs. Gram-negative bacterial cells.

3. What is the differential step in the Gram staining process? Explain your answer.

4. What staining result would you expect if you prepared a Gram stain from a culture containing both Gram positive cocci and Gram negative bacilli and you forgot to add the counterstain? Describe the results for both Gram-positive cells and Gram-negative cells

Practical Exercise 6A: Gram Stain and Microscopy

1. You will be assigned a bacterial culture to Gram stain.

2. After you have prepared a Gram stain, you will be asked by your TA to set up your slide for viewing on a designated microscope.

3. The TA will observe your actions as you work and will grade you on your microscope technique and the quality of your Gram stain. (Don't forget to clean up the microscope for the next student when you are finished.)

Section II: Transmission of Disease and Control of Microbial Growth

Understanding how diseases are spread is the first step in understanding how disease transmission can be controlled. The factors involved in transmission of a particular pathogen include characteristics of the organism such as the sites of entry into and exit from the body, how long the pathogen can survive outside the body and the possible existence of an environmental reservoir of the organism. Several exercises are included in this section to demonstrate the ease of transmission of pathogenic organisms.

Chemical and physical methods to combat the spread of various pathogens: Gram positive and negative bacteria, spore formers, acid fast bacilli, etc. have been developed. Experiments included in this section demonstrate some of these techniques and relate the effectiveness of various control methods to the characteristics of particular organisms

Exercise 7: Transmission of Disease

Objective: To study several modes of disease transmission and improve understanding of how to control the spread of disease. Upon completion of this exercise you should be able to:

❖ Define and give an example for each type of transmission discussed in this exercise.

Application: In the clinical setting and in epidemic situations, diseases are spread by direct, indirect, or vehicle transmission. It is only possible to stop the chain of disease transmission when you understand how it is spread.

A communicable disease by definition is a disease that can be transmitted. Hence, the primary focus in the study of epidemics (epidemiology), both large and small, is how a particular disease is transmitted. In order to prevent, or stop, an epidemic you must first know how the disease is spread and then find a way to intervene.

One aspect of characterizing a disease is its mode of transmission, a characteristic directly related to the etiological agent and the site of infection. For example, respiratory diseases are usually spread directly from one human to another by droplet transmission because infectious microbes are present in secretions of the respiratory tract. However, if the microbe being transmitted can live outside the body, any person coming into contact with the source can contract the infection

Modes of Transmission

Direct Transmission: Transmission of disease directly from one host to another.

Droplet transmission occurs when bacteria adhere to very tiny particles of saliva/mucus that are emitted from the respiratory tract upon sneezing, coughing, or talking. These tiny droplets can be inhaled by another human. If sufficient numbers of organisms are inhaled, an infection may ensue.

Contact Transmission: Transmission of disease by intimate contact, e.g. touching, kissing, etc.

Indirect Transmission: Host to host transmission involving inanimate objects or vectors.

Fomites are inanimate objects that may/can transmit disease. The shared use of a drinking glass, towel, pencil, money, etc could result in transmission of disease from an infected person to others via the inanimate object.

Vector Transmission involves another organism. Usually the organism (called the vector) is an arthropod such as a tick, flea, or mosquito, which takes a blood meal from an infected individual and upon feeding again, transmits the microbe to a new host. In some cases the microbe is able to live in the vector and the vector acts as an environmental reservoir for the disease. In other cases the pathogen is only carried by the vector, as for example organisms that are transferred on the appendages of flies.

Vehicle Transmission: Transmission via ingestion or injection of contaminated substances.

Contaminated Food or Water is often a vehicle for disease transmission. Ingestion of such substances can spread disease to many hosts in a very short period of time.

Contaminated Blood Products injected into an otherwise healthy person can very quickly cause serious illness.

You will carry out three exercises designed to demonstrate how diseases can be transmitted.

Exercise 7a: Aerosols - Droplet Transmission

Sneezing, coughing, and talking are important mechanisms for direct contact transmission of infectious diseases. We produce aerosols, tiny airborne droplets, during sneezing, coughing, and talking. These droplets vary in size from macro- to microscopic. The smaller droplets are less than 10 m in diameter and may contain a couple of bacterial cells or thousands of viral particles. These microorganisms when inhaled are small enough to enter the lung. In a sneeze, enormous numbers of droplets are released and can travel at about 100 meters/second (over 200 MPH). Coughing and shouting release droplets traveling at about 16-48 meters/second. These tiny drops quickly evaporate and leave behind a nucleus of mucus to which bacteria are attached. Aerosols are the main method of transmitting respiratory diseases and account for their contagious nature.

Procedure:

1) Obtain 2 nutrient agar plates (NAP)

2) Plate #1
   a. Remove the lid, hold the agar surface in front of your face approximately 8 inches away from your mouth and cough vigorously.
   b. Incubate the plate at 37° for 24 hours.
   c. Count the colonies and the colony types on your plate.
   d. Determine the Gram reaction of the predominant colony type.
   e. Record your results in **Table 7a** in the **Results** section.

3) Plate #2
   a. Hold the agar surface about 8 inches in front of your mouth, wet your lips as you would when preparing to speak, and recite "Peter Piper Picked a Peck of Pickled Peppers". Project your voice as if you are speaking to a room full of people.
   b. Incubate the plate at 37° for 24 hours.
   c. Count the colonies and the colony types on your plate.
   d. Determine the Gram reaction of the predominant colony type.
   e. Record your results in **Table 7a** in the **Results** section.

Exercise 7b: The Handshake - Direct Contact Transmission

Direct transmission via touching is a common means of disease transmission. For example, Hepatitis A virus and several other infectious diseases the major route of transmission is via the hands of food handlers. The importance of effective and frequent handwashing, especially among hospital workers and food handlers, cannot be over emphasized.

The objective of this experiment is to form a sequential transmission cycle. Each person will be assigned a number in the sequence and each student will be given lotion to apply to their hands. Each will proceed to shake the hand of the person before and then after them in the numerical sequence. One unidentified student in the sequence will be issued fluorescent lotion; the others will be issued regular lotion. The results will show where the contamination occurred in the sequence and will demonstrate how far bacteria can be spread from one contaminated hand to others. It will become very obvious that handwashing can be a very effective means of stopping the transmission of disease.

Procedure:

1. Each person will be assigned a number in sequence. (1, 2, 3, etc.).

2. Each person will be given a small amount of lotion which they will apply to their right hand.

3. The person with the lowest number will begin a hand shake sequence by shaking hands (using the right hand) with the person holding the next highest number (i.e. 1 shakes hands with 2). The second person will then shake hands with number three and then 3 with number 4 and so on until the person with the highest number is reached. The person with the highest number will then shake hands with the person assigned number 1. At this point each participant will have shaken hands with two people- the person before and the person after them in the numerical sequence.

4. The TA will then use a small black light to show you who had the fluorescent lotion and how far it has traveled just by shaking hands.

5. Follow the instructions provided in the **Results section 7b** and record your observations.

Exercise 7c: Fomites - Indirect Contact Transmission

Individuals with an infectious disease may shed pathogens into their environment by direct contact with objects or by releasing aerosols that settle on surrounding surfaces. While direct contact with an infected individual is an obvious mechanism for disease transmission, contact with a contaminated fomite (inanimate object) is another important mode of transmission (indirect contact). Fomites do not support the growth of microorganisms; microorganisms normally survive on them for only a matter of hours. However, during that interval several individuals may come into contact with the contaminated fomite and become infected. Similarly, an individual who contacts a contaminated object may transmit the infection to numerous people.

Procedure:

1. Lay two paper towels on your workbench.

2. Obtain a white and a blue spoon from your TA and put each on a separate paper towel.

3. Handle the white spoon as you would if you were going to use it. Hold it and move it around in your hand for about 30 seconds and then put it back on its paper towel.

4. Pick up the blue spoon with the same hand you used to handle the white spoon. Again, handle it as though you were using it and move it around in your hand for about 30 seconds and then it put back on the paper towel.

5. The TA will now show you what would happen if the white spoon you handled first was contaminated with a pathogen by showing that a fluorescent compound contaminating the white spoon was transferred to your hand and the blue spoon.

6. Record your observations in section **7c** of the **Results** section.

Results

Table7a Record the results you obtained from the droplet transmission experiments.

	# colonies/plate	# colony types/plate	Gram Stain Reaction of Predominant Colony Type
Cough Plate			
Talk Plate			

7b- The Handshake

Make a table and record the class results for the handshake experiment. Record the amount of fluorescence observed on each student's hand.

No fluorescence = (-); Little = (+); Moderate = (++); Heavy = (+++)

Interpret the results of the handshake experiment.

7c- Fomites

Interpret the results of the fomite experiment. What did you observe when you put your hand under the UV light? What did you observe when you put the blue spoon under the UV light?

Questions

1. Differentiate between direct and indirect transmission. Give an example of each using a situation that could occur in your chosen occupation.

2. Define and differentiate vector transmission and vehicle transmission. Give an example of each.

Exercise 8: Bench Survival

Objectives: To determine how long organisms survive when held in a dry state, at room temperature, with no nutrients. Upon completion of this exercise you should be able to:

- ❖ Define endospore.
- ❖ Describe the characteristics of an "acid-fast" cell wall.
- ❖ State how long each of the bacteria tested could survive at room temperature without nutrients or water.
- ❖ Explain why some of the bacteria survived desiccation better than others.
- ❖ Explain why this knowledge is important in a clinical or food handling setting.

Application: Especially pertinent with nosocomial diseases, it is important to understand that certain bacteria can live for long periods outside the body, on inanimate objects.

Most vegetative (actively growing) bacterial cells cannot survive desiccation (drying out) or lack of nutrients for long periods of time. For organisms that produce them, such as *Bacillus* sp. (primarily soil organisms), endospores are a means of survival when adverse conditions such as drought cause the soil to dry out. Endospores, essentially a dormant state in the cell life cycle, are only produced by certain genera of bacteria. In the process of endospore formation, the cell makes a new copy of its DNA and surrounds it with membranes and a spore coat that is well protected from the environment: impervious to many chemicals, waterproof so the endospore cannot dry out, and resists high temperatures without breaking down. Once the endospore is triggered to grow again (germination) by favorable conditions, the spore coat breaks away, the intact copy of the DNA allows proteins to be made again, metabolism resumes, and the vegetative cells begins to divide and grow again. Bacteria can survive in the form of endospores for years; for example, viable endospores have been isolated from ancient tombs that are known to be thousands of years old.

Acid-fast cells, notably the genus *Mycobacterium* (genus of tuberculosis and leprosy), are known for their survival under adverse conditions although they do not have the longevity of endospores. To be designated acid-fast, a vegetative cell has very high lipid content in the cell wall, requiring the use of heat or other means to drive the stain into the cell and once in, even acid alcohol cannot decolorize it. These lipids are large, bulky, possess some of the characteristics of waxes, and make the cells somewhat waterproof. Due to these characteristics, the cells are resistant to desiccation and other adversity and can survive without water or nutrients for days on inanimate objects.

In hospitals, careless personnel may inadvertently carry infectious disease from one patient to another. Many of the principles of patient care were developed to prevent this transmission of disease. For most diseases the procedures are adequate. However, in the case of microbes that can survive desiccation for long periods, special precautions are necessary. *Mycobacterium tuberculosis* patients, for example, must be under strict isolation; their entire room, including furniture, equipment and devices must be disinfected upon release of the patient.

Procedure:

The four organisms listed below will be used to demonstrate the ability of some organisms to survive desiccation on inanimate objects. Note: The organisms you will be testing in this exercise are not the strict pathogens but their resistance properties are similar to that of the true pathogens in the respective genera.

1. *Mycobacterium smegmatis* - acid fast
2. *Bacillus cereus*- Gram positive bacillus- sporeformer
3. *Escherichia coli*- Gram negative bacillus
4. *Staphylococcus aureus*- Gram positive cocci

Lab 1

1. Divide an empty plate into quarters and label them 1-4. Draw a nickel-sized circle in the center of each segment

2. Using a cotton swab, make a smear of *Mycobacterium smegmatis* within the circle in quadrant one. Repeat with *Bacillus cereus* in segment 2, *Escherichia coli* in segment 3 and *Staphylococcus aureus* in segment 4. Allow the smears to dry.

3. When the smears are dry, swab one of the smears and subculture the organisms to a nutrient agar plate. Use the technique described below to produce a "lawn" of bacterial growth.
   a. Gently brush the swab over the entire surface of the agar.
   b. Rotate the plate $90°$ and swab the entire surface of the agar in a direction perpendicular to the first inoculum.
   c. Finish this process by swabbing the agar at a $45°$ angle to the first inoculum.

4. As described in step 3, subculture the 3 other organisms to nutrient agar plates.

5. Save the dry smears you prepared. .

6. Incubate the plates at $37°$ for 24-48 hours. Note: *Mycobacterium smegmatis* may take a few days to grow up when it is prepared from a dried smear. *Mycobacterium* plates should be removed from the incubator after 48 hours and placed at room temperature for an additional 2 days.

Lab 2

1. After seven days, to determine whether organisms in the dry smears you prepared previously (step 2 above) are still viable, subculture the smears following the procedure described in step 3 above.

2. A set of demonstration plates will be available for examination. The organisms on these plates were obtained by swabbing dry smears daily for seven days. This set of plates represents the results you would have obtained if you had sampled at daily intervals the dry smears you prepared and shows approximately how long each organism survived.

3. Observe your plates and the demonstration plates. Estimate the length of time each of the four test organisms survived on the inanimate object.

4. Record your results in **Tables 8.1** and **8.2**.

Results

Table 8.1 Survival of test organism on demonstration plates. The symbol (-) represents no growth, (+) light growth, (++) moderate growth; and (+++) heavy growth.

	Day 1	Day 2	Day 3	Day 4	Day 5	Day 6	Day 7
Mycobacterium smegmatis							
Bacillus cereus							
Escherichia coli							
Staphylococcus aureus							

Table 8.2 Survival of the organisms you tested. The symbol (-) represents no growth, (+) light growth, (++) moderate growth; and (+++) heavy growth.

	Day 1	Day 7	Comments
Mycobacterium smegmatis			
Bacillus cereus			
Escherichia coli			
Staphylococcus aureus			

Questions

1. List each bacterium used in this experiment. State the property each has that differentiates it from the others and how this property relates to the expected survival outcome?

2. What are the physical properties that characterize an acid-fast bacterium?

3. What is the function of endospore formation?

Exercise 9: Effectiveness of Antiseptics and Disinfectants

Objective: To understand the effects of various disinfectants and antiseptics and to test their effectiveness against several common bacteria. Upon completion of this exercise you should be able to:

- ❖ Define and differentiate antiseptics and disinfectants.
- ❖ List factors which determine the effectiveness of antiseptics and disinfectants.
- ❖ Define CT value.
- ❖ Calculate the CT value of a chemical.
- ❖ Interpret relative effectiveness of antiseptics/disinfectants when given their CT value.

Application: Disinfectants and antiseptics are used everyday of our lives. Several parameters affect the use of chemical disinfectants; awareness of these parameters can increase effective use of the disinfectants.

Disinfectants are chemical substances used to kill microorganisms on inanimate objects. Antiseptics are chemical substances used to kill microorganisms on skin and mucus membranes. The definitions dictate that antiseptics are usually less effective killing agents than disinfectants because they must be mild enough to cause no harm to the skin or membranes. Often the chemical substances are the same and it is only the concentration that is varied to determine if it can be used as an antiseptic.

Another factor that affects the killing effectiveness of disinfectants is time. The longer a microorganism is exposed to a disinfectant, the more effective it will be. So in the case of antiseptics, where the concentration is limited, the time of exposure can be increased to increase the kill rate.

When disinfectants and antiseptics are developed, time and concentration are the parameters used to evaluate their effectiveness. To facilitate the comparison of various chemicals, regulatory agencies establish what is called a **CT value** for the chemical. The CT value is determined by multiplying the concentration of the chemical times the length of contact (in minutes). The CT value is determined by the manufacturer and it is compared to the disinfectant, phenol, the "gold standard" of disinfectants. It must be determined under standard conditions for each organism. For example, the $CT^{chlorine}$ value for *E. coli* is 0.5 and for *Legionella pneumophila*, 1200. Therefore, we can conclude from these data that *E. coli* is very susceptible to low concentrations of chlorine for short periods of time while *Legionella* is 2400 times more resistant to chlorine.

In most areas of medicine, disinfectants and antiseptics are used routinely to fight infectious agents. Good disinfectants have low CT values for common pathogens including viruses and fungi. The best disinfectants also have low CT values for bacterial endospores as well as *Mycobacterium*.

Procedure:

1. Obtain one tube of each of the solutions (disinfectant or antiseptic provided at their working dilution) to be tested.

2. Obtain an overnight culture of the bacterium to be used for the tests.

3. Add 0.1 ml of the overnight culture to each tube of disinfectant/antiseptic solution and mix well.

4. Incubate at room temperature. After 1, 7, and 15 minutes transfer 0.1 ml of the test sample to a nutrient agar plate. Use a sterile hockey stick to spread the sample over the surface of the agar.

5. Incubate the plates at 37°C for 18-24 hours and observe. If the plates contain less than 300 colonies, count the colonies. The plate counts will allow you to determine the number of viable bacteria per ml in the disinfectant/antiseptic solution at the end of each incubation period.

6. Record the results in **Table 9.1** and **Table 9.2**.

Results

Table 9.1 Record class results in the table below. Enter the number of viable organisms per ml remaining in the disinfectant after the incubation time.

	Disinfectant	1 min.	7 min.	15 min.
Escherichia coli				
Staphylococcus aureus				
Bacillus cereus				

Table 9.2 Record class results in the table below. Enter the number of viable organisms per ml remaining in the antiseptic after the incubation time.

	Antiseptic	1 min.	7 min.	15 min.
Escherichia coli				
Staphylococcus aureus				
Bacillus cereus				

Questions

1. Assume you that you started with 1×10^8 cells per ml and calculate the efficiency of the kill for each disinfectant/antiseptic. (Efficiency is the % of total organisms that were killed by the disinfectant/antiseptic.)

2. What is the difference between a disinfectant and an antiseptic?

3. What factor(s) determine the effectiveness of a disinfectant/antiseptic?

Exercise 10: Effectiveness of Handwashing

Objective: To observe the effectiveness of handwashing for removal of microbes from the hands. Upon completion of this exercise you should be able to:

- ❖ Define and differentiate normal flora and transient microbes.
- ❖ Explain what happens during each step of handwashing.
- ❖ Explain the differences between the different methods of handwashing

Application: Handwashing is strongly emphasized with laboratory and hospital personnel and food handlers as an effective means of preventing the spread of pathogens. Surveillance has shown that even a seemingly simple task such as handwashing, if properly performed, does prevent the transmission of disease.

The human body is home to millions of microorganisms. The skin of the developing fetus is sterile; after birth we are colonized, both inside and out, by microbes. These microbes constitute what is referred to as normal flora. Normal flora consists of bacteria that are benign, often beneficial, and sometimes essential for proper body function. In contrast to normal flora, transient microbes are bacteria that can persist on or in us for only a few days, weeks, or months. Transient microbes can be opportunistic or pathogenic.

One of the earliest efforts to control infections resulted from the recognition in the 1800's that women were dying after childbirth from blood stream infections. Investigation revealed that infections were being spread because physicians did not wash their hands after examining each patient.

The numerous microbes residing on the hands consist of both normal flora and transient microbes. Hand washing is still the cornerstone of modern infection control programs. Appropriate hand washing procedures involve the use of soap to remove oil and scrubbing to help remove bacteria that lie beneath dead skin cells. Today, hand sanitizers, most of which contain 62% ethyl alcohol as the active ingredient, are popular because they can be used to disinfect the hands when soap and water are not available.

In this exercise you will investigate the effectiveness of hand washing with different products to remove and/or kill bacteria. Since the organisms you will culture are only those which grow aerobically on nutrient agar, organisms with different nutritional and environmental requirements will not grow. However, many normal skin inhabitants will grow.

Procedure:

Lab 1

1. Obtain a nutrient agar plate and label one half "before" and the other half "after".

2. Moisten a sterile swab with sterile water.

3. Swab your nondominant hand as follows: on the palmar surface of the hand begin at the top of the index finger and swab down to the base of the thumb. Roll the swab and swab back up to the fingertip of the index finger. Use the same swab and repeat the procedure but swab the palmar surface of the ring finger. Use this swab to inoculate the "before" side of the nutrient agar plate.

4. You will be assigned one of the following hand washing procedures:
   a. Wash your hands with hot water only
   b. Wash your hands with soap and hot water.
   c. Wash your hands with soap and hot water and in addition use a brush to thoroughly scrub your hands.
   d. Use the hand sanitizer as per directions on the bottle.

5. After washing your hands, swab your nondominant hand with a moistened swab as follows: on the palmar surface of the hand begin at the top of your second finger and swab down to the base of your palm. Roll the swab and swab back up to the fingertip. Using the same swab, repeat this process on the palmar surface of the fourth (pinky) finger. Use this swab to inoculate the "after" side of the nutrient agar plate.

6. Incubate the plate at 37° for 24 hours.

Lab 2

1. Observe any differences between the two sides of the nutrient agar plate.

2. Record your results in **Table 10.1**. Include in your results the class data obtained following each of the washing procedures.

Results

Table 10.1 Estimated growth of cultures from hands before and after various washing procedures. None= (-); Light = (+); Moderate = (++); Heavy = (+++).

	Before washing	After washing
Water only		
Soap and water		
Scrub		
Sanitizer		

Interpret the results. Which method most reduces the amount of growth? Explain why. Were there any anomalies (e.g. some people had more microbial growth after washing than before washing)? Speculate as to why these may have occurred.

Questions

1. What results did you <u>expect</u>? Explain your reasoning.

2. Define normal flora. Name three sites on the body where you find normal flora and state how it is beneficial.

3. Define transient microorganism. Give an example.

4. What effect does soap have when used during handwashing? Scrubbing?

Exercise 11: Antibiotic Susceptibility Testing

Objective: To understand the principles of antibiotic susceptibility testing. Upon the completion of this exercise you should be able to:

> ❖ Define antibiotic.
> ❖ Differentiate between broad-spectrum and narrow-spectrum antibiotics.
> ❖ List 2 factors contributing to the development of antibiotic-resistant stains of bacteria.
> ❖ List 2 types of mechanisms of action of antibiotics.
> ❖ Explain the importance of testing antibiotic susceptibility.
> ❖ Define and interpret zone of inhibition.
> ❖ Explain why it is important to start with a pure culture when performing an antibiotic susceptibility test.

Application: Routinely all bacterial patient isolates are tested for their susceptibility to antibiotics. Critical levels of antibiotic resistance are in part due to the fact that bacteria can share the genes for resistance by sharing small pieces of their DNA, called plasmids. The effect of having a resistance plasmid, or R-factor, will be demonstrated.

Since Alexander Fleming's discovery of penicillin in 1929, antibiotics have become the standard method used to treat bacterial diseases. Antibiotics are chemical compounds that selectively inhibit or kill microorganisms while causing little or no damage to animal cells. For this reason, they can be introduced into the human body to treat bacterial infections with minimal effects on human cells. Because of occasional adverse side effects and the increasing emergence of antibiotic resistant organisms, antibiotics must be used properly and judiciously to safeguard their clinical usefulness.

Table 11.1 lists some common antibiotics, their effect on bacterial cells, and their spectrum of activity. Some affect cell wall production of bacteria, some inhibit protein synthesis in the bacterial cell, and others disrupt cell membranes, etc. A broad-spectrum antibiotic is effective against a wide variety of bacteria and usually affects an aspect of the cell such as protein synthesis that is common to all cells. Narrow-spectrum drugs are effective against specific bacteria, usually targeting either Gram-positive or Gram-negative bacteria. .

Bacteria are especially prone to developing resistance to the various antimicrobial agents. There are 3 major mechanisms of antimicrobial resistance that have developed in various bacterial populations. These are: 1) inactivation of the drug by the bacterium; 2) blocking or mutation of the active site of drug action in the bacterial cells; and 3) preventing the drug from entering the cell and reaching the active site of drug action. To become resistant, bacteria often have to make only a single resistance protein. The single gene required to synthesize such a protein is often encoded on a plasmid (a small piece of DNA in addition to the chromosome). Plasmid copies can be transferred to neighboring cells rendering the recipient resistant to the drug also. Thus, a mutation in a single cell that results in drug resistance can quickly become a characteristic common to millions of cells. The result is an antibiotic resistant strain.

When a disease-causing bacterium is isolated from a patient, the antibiotics that will successfully treat the infection must be determined. One of the oldest methods for evaluating the effectiveness of a particular antibiotic against a specific type of bacteria is the Kirby-Bauer method. In this method, Mueller-Hinton agar is inoculated with a pure culture of the organism being tested. Filter paper disks, impregnated with antibiotic, are then placed on the agar surface. While the plates are being incubated to promote bacterial growth, the antibiotic diffuses through the medium. After incubation the plates are inspected. If the antibiotic is effective, bacterial growth around the disk is inhibited as indicated by the clear zone. The diameter (in millimeters) of the zone of inhibition is measured and compared to a standard to determine if the bacterial isolate is susceptible, resistant, or intermediately resistant to the drug.

Table 11.1 Summary of some commonly used antibiotics

Antibiotic	Mechanism of Action	Spectrum of Activity
Ampicillin	Inhibits cell wall synthesis	Broad; Gram-positive and Gram-negative bacteria
Bacitracin	Inhibits cell wall synthesis	Narrow; Gram-positive
Chloramphenicol	Inhibits protein synthesis	Broad; Gram-positive and Gram-negative bacteria
Erythromycin	Inhibits protein synthesis	Narrow; Gram-positive
Gentamicin	Inhibits protein synthesis	Broad; Gram-positive and Gram-negative bacteria
Penicillin G	Inhibits cell wall synthesis	Narrow; Gram-positive cocci
Polymyxin B	Disrupts the cell membrane	Narrow; Gram-negative rods
Rifampin	Inhibits RNA synthesis	Broad; Gram-positive and certain Gram-negative (e.g. *N. meningitidis*)
Streptomycin	Inhibits protein synthesis	Broad; Gram-positive and Gram-negative bacteria
Tetracycline	Inhibits protein synthesis	Broad; Gram-positive and Gram-negative bacteria
Vancomycin	Inhibits cell wall synthesis	Narrow; Gram-positive

Table 11.2 Zones of inhibition for antibiotics.

Antibiotic (concentration)	Diameter of zone of inhibition (mm)		
	Resistant $\leq$	Intermediate	Susceptible $\geq$
Ampicillin (10 g)	13	14-16	17
Bacitracin (10 units)	8	9-12	13
Chloramphenicol (30 g)	12	13-17	18
Erythromycin (15 g)	13	14-22	23
Gentamicin (10 g)	12	13-14	15
Penicillin G (10 units)	11	12-21	22
Polymyxin B (300units)	8	9-11	12
Streptomycin (10 g)	11	12-14	15
Tetracycline (30 g)	14	15-18	19
Vancomycin (30 g)	9	10-11	12

Procedure:

Lab 1: Inoculation and preparation of plates

1. In this exercise you will determine the susceptibility of a strain of *Escherichia coli* carrying a resistance plasmid, *Escherichia coli,* and *Pseudomonas aeruginosa* to various antibiotics.

2. Make a suspension of the bacterium to be tested. For this procedure you will need the following:
   a. a bacterial culture provided on an agar medium (slant or plate)
   b. a water blank
   c. a 0.5 McFarland standard (the standard contains an insoluble barium salt that serves as a turbidity standard).

   Make the bacterial suspension by adding bacteria from the agar culture to the water blank until the turbidity approximates that of the McFarland standard.

3. Use a sterile swab to inoculate a Mueller-Hinton Agar plate with the suspension of bacteria. Follow the techniques described below to produce a "lawn" of bacterial growth which is critical in this exercise:
   a. Use a swab to gently brush the bacterial suspension over the entire surface of the agar.
   b. Rotate the plate 90° and swab the entire surface of the agar in a direction perpendicular to the first inoculum.
   c. Finish this process by swabbing the agar at a 45° angle to the first inoculum.

4. Using sterile forceps pick up an antibiotic disk from one of the sterile Petri dishes and place it on the agar surface. Position the disk approximately 3cm from the edge of the plate. Press the disk **very gently** onto the agar using the tip of the forceps.

5. Repeat step 4 for each antibiotic disk to be tested. To ensure accurate results, position the 6 disks on the plate according to the pattern illustrated below. Be sure to correctly identify the antibiotic associated with each disk you have placed on the plate.

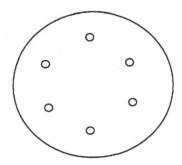

6. Repeat steps 2 to 5 but with a different organism.

7. Incubate the plates for 18-24 hrs at 37°C.

Lab 2: Examination of plates

1. Observe the plates for presence or absence of growth around each antibiotic disk.

2. Use a ruler to measure the diameters of the observed zones of inhibition around each antibiotic disk.

3. Using **Table 11.2** as a reference, indicate whether the organism is resistant, intermediate, or susceptible to each antibiotic.

4. Record your results in the **Table 11.3**.

Results

Table 11.3 In column 1, list the antibiotics tested. For each organism give the Kirby Bauer result including whether it is resistant, intermediate or susceptible. This information is available in **Table 11.2**.

Antibiotics	*Pseudomonas aeruginosa*	*Escherichia coli*	*Escherichia coli* with resistance plasmid

Interpret the results of the antibiograms for these organisms. Which antibiotic was most effective against each of the 3 microbes? Describe the differences in antibiotic resistance between the *E. coli* and the *E. coli* with a plasmid for resistance. Were they what you expected?

Questions

1. Define antibiotic.

2. List two factors responsible for the emergence of antibiotic resistant strains of bacteria. (Refer to http://www.hhs.gov/news/press/2001pres/01fsdrugresistance.html and/or www.cdc.gov for further information)

3. List 2 mechanisms of action of antibiotics.

4. Define a narrow-spectrum antimicrobial. Give an example.

5. When performing the Kirby-Bauer antibiotic susceptibility test, why is it critical to start with a pure bacterial culture?

6. What is a zone of inhibition?

Section III: Basic Clinical Microbiology

Clinical Microbiology is the study of microorganisms that infect the human body. These microorganisms can be categorized into two types: the true pathogens and the opportunists. The true pathogens are those that, if allowed to enter the body in sufficient numbers, will cause disease. The opportunists are often normal flora in certain parts of the body and when displaced into a suitable area they can set up an infection.

The true pathogens are those that cause what we often view as the classical diseases: gonorrhea, syphilis, AIDS, anthrax, tuberculosis, etc. Any time these organisms are allowed to enter the body and grow they cause damage to the host. They are never found in healthy individuals.

Opportunists are just that; microbes that often reside in the body at various non-sterile sites in a non-pathogenic state but infect when the opportunity arises. However, this does not mean that they are lesser pathogens for if allowed to enter sterile regions of the body they can set up damaging infections and sometimes kill their host just as if they were true pathogens.

The proper collection of clinical specimens is critical to the patient's outcome. In most cases, infected sites contain only one or possibly two species. This is so because competition between organisms for resources is great and in most cases one organism will have the strength and virulence to outgrow all of the others. This knowledge has application in the clinical setting. For example, if midstream urine is cultured and numerous types of bacteria grow from the specimen, the trained microbiologist will recognize that such a mixed culture is indicative of contamination. A new sample will be requested before results are released and the patient may have to wait for another culture before treatment is started. Obtaining a clean, valid specimen from an infected site can determine if the patient is promptly treated and treated with the correct antibiotic.

Transport of specimens is another critical area where understanding and training in proper techniques can ensure the patient will receive proper treatment. When fastidious organisms are collected, time and proper transport media can determine if the organism survives to be identified in the laboratory and therefore whether the patient receives the correct diagnosis. On the other hand, when specimens with normal flora are allowed to incubate at room temperature, the normal flora can grow to high numbers and be misinterpreted as infecting organisms. In this case the patient may be treated for an infection that does not exist.

As is evident from the above examples, timely processing of the specimens and accurate interpretation of the results by the clinical microbiologist is dependent upon proper collection and transport of patient specimens to the laboratory. These factors will make the difference in whether the patient receives proper, timely care.

Exercise 12: Transport of Specimens

Objective: To develop the understanding that patient specimens must be handled with care to ensure the quality of the specimen when it reaches the laboratory for processing. Upon completion of this exercise you should be able to:

❖ Define generation time.
❖ List and understand the significance of the three specified considerations regarding specimen collection.
❖ Describe the purpose of the Bio-Bag system.
❖ Explain how leaving a clinical specimen at room temperature will affect the diagnosis of infection.

Application: Many microorganisms are difficult to culture from the host. Often, special collection procedures must be followed to ensure survival of the pathogens on artificial media. Awareness of these methods is imperative when working with patients or in the medical laboratory.

The generation time of most pathogens is approximately 30 minutes. This means that cells double in number about every 30 minutes. Growth of bacteria on a nutrient agar plate is rapid. A single cell can divide to form a colony containing thousands of cells in 18-24 hours. In body fluids bacteria can do the same. The collection and delivery of patient specimens to the lab for processing are primary responsibilities of medical personnel. If this is part of your job, it is imperative that you use your understanding of bacterial growth when performing this task.

The very special requirements for the proper collection and transport of microbiological specimens take into account three important considerations:

1. Often the number of organisms cultured from body fluid is the determining factor in deciding whether the organism is infecting the patient or whether it is normal flora. Consequently, patient specimens must be handled properly to ensure accurate laboratory results and therefore proper treatment for the patient. You learned from your study of the relationship between growth and temperature that all organisms have an optimum temperature for growth. Human pathogens usually grow best at body temperature (37°C); temperature outside the body (room temperature, 21-22°C) is sub-optimal. At room temperature growth is slowed down but not stopped. If you lower the temperature even more (refrigeration, 4°C) growth is slowed even more. Thus most specimens can be stored in the refrigerator overnight without affecting the laboratory results.

2. Proper sterile technique when collecting specimens is also imperative. Many sites in the body are normally sterile and so the presence of a single organism in a culture results in the prescription of antibiotic therapy for the patient.

3. Some bacteria and most viruses die if stored for extended periods outside the body. For the more fastidious organisms, special transport media must be used to ensure that the organism survives transport to the lab.

Figure 12.1 Illustrates several types of containers used to collect and transport patient specimens. Some of the containers such as the tubes and bottles shown on the left contain special media that ensures the survival of bacteria and viruses for relatively short periods of time while the specimen is in transit to the laboratory.

Figure 12.2 Shows one such transport device in greater detail. It consists of a sterile swab, shown at the top of the illustration, which is used to collect an aerobic specimen. After the specimen is collected, the swab is placed in the sheath shown at the bottom of the illustration. A gel-like medium fills the bottom third of the tube. The medium contains a buffered saline solution that keeps organisms moist and at proper pH and salinity during transport to the lab. The medium is not a growth medium and therefore the culture should be promptly delivered to the laboratory for plating on nutrient media.

Many different transport media are available; it is important that the clinician consult the laboratory to determine what transport device is appropriate for the specimen they wish to collect. Laboratories supply these transport devices and include the cost as part of performing the culture. Also, laboratories often provide literature describing the particular transport devices appropriate for obtaining particular cultures Clinicians are encouraged to contact the laboratory of there choice for literature on transporter use.

Figure 12.1 Various transport media

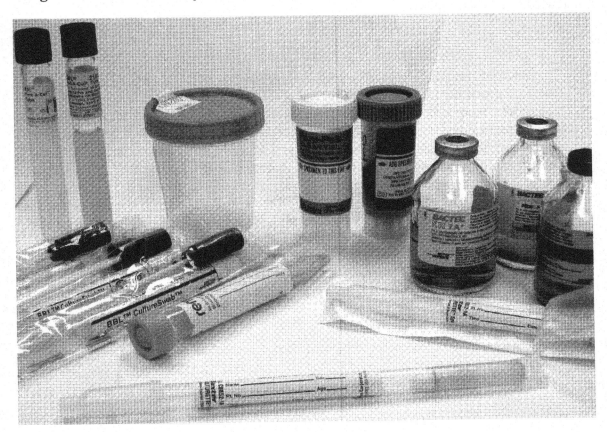

Figure 12.2 Commonly used transporter for aerobic culture

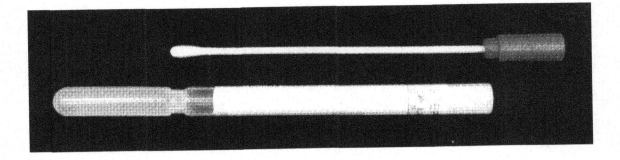

Neisseria gonorrhoeae is one of the more fastidious pathogens. Special media for growth and special means for transport must be used with this organism. To successfully culture *N. gonorrhoeae*, a collected specimen must be immediately transferred to Thayer-Martin medium and an increased CO_2 environment.

Procedure:

1. Thayer-Martin media is used to isolate *N. gonorrhoeae*. It is a very rich medium containing several antibiotics that allow the selection of *N. gonorrhoeae*. The culture you will be using is a non-pathogenic strain of *Neisseria lactamica,* an organism with growth requirements similar to *N. gonorrhoeae*. The media you will be using is chocolate agar. Chocolate agar is similar to Thayer-Martin agar but lacks the selective antibiotics. Streak the chocolate plate for isolation with the *N. lactamica.*

2. Place the Petri dish and the CO_2 generator in the Bio-Bag and seal the bag. Hold the generator upright and crush the vial. Bubbling should occur. If not, gently flick the lower portion of the vial with your finger. The vial must remain upright for 30 sec.

3. Incubate the Bio-Bag at 37°C for 18-24 hrs and observe growth. Colony morphology of *N. lactamica* on chocolate agar is very similar to that of *N. gonorrhoeae* on Thayer-Martin agar.

4. Perform a Gram stain on one of the colonies.

Results

1. Give colony morphology of the organism from the chocolate agar plate. If there is no growth on your plate, give a plausible reason for why it did not grow.

2. Draw the Gram stain result.

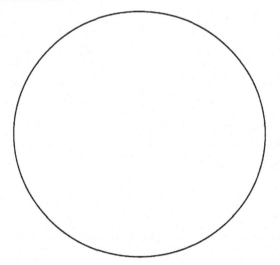

Questions

1. Describe the purpose of the Bio-Bag transport system.

2. Give two examples of critical mistakes which could occur during the collection, storage, or transport of patient specimens. Include in your answer the possible outcome/effect of these mistakes on the specimen, or ultimately on the patient.

Exercise 13: Selective and Differential Media

Objective: To understand the theory behind selective and differential media and to learn some of the uses of these special media in the clinical microbiology laboratory. Upon completion of this exercise you should be able to:

- ❖ Define and give an example of a selective medium.
- ❖ Define and give an example of a differential medium.

Application: In most cases, the initial isolation media used in the clinical setting is a selective and/or differential media. The purpose for this is to promote the isolation of the suspected pathogens by inhibiting the growth of many of the normal flora.

The media used in the exercises thus far have been general undefined growth media that supply the nutritional needs of most pathogens. If an organism is resistant to an antibiotic that inhibits other flora, the drug can be added to the growth medium to select for the growth of the organism of interest. This is called a selective medium. A differential medium, by contrast, contains a substance(s) that if used by the organism causes a visible change in the medium.

Mannitol Salt Agar (MSA) is a good example of both a differential and selective medium. This agar contains mannitol and a pH indicator that changes color when mannitol is fermented and thus acid is produced. The salt concentration is 7.5%. Of the normal human flora, only *Staphylococcus* sp. can grow in 7.5% salt. Therefore, MSA is selective for *Staphylococcus* and differential because a change in the color of the pH indicator from red to yellow indicates the organism is able to ferment mannitol.

Eosin Methylene Blue (EMB), Hektoen Enteric and MacConkey agars contain substances that inhibit the growth of Gram positive organisms. Further, all three media differentiate between lactose fermenting and non-lactose fermenting organisms. Many of the common intestinal pathogens cannot ferment lactose; therefore the existence of non-lactose fermenting colonies on plated stool samples allows one to immediately differentiate potential pathogens from normal fecal flora.

Blood agar is an example of a differential media that is not selective. However, blood agar can be used to distinguish many pathogens, especially members of *Streptococcus* and *Staphylococcus* groups on the basis of their hemolytic properties (β, α, and γ).

Demonstration:

1. Several types of selective and/or differential media will be available after inoculation with appropriate organisms to demonstrate the various properties of each medium.

2. View each medium and record the organisms and the reactions.

3. Using your textbook, research the selective and differential qualities of each medium. Determine the chemicals that give the medium its selective or differential quality.

Results

Table 13.1 Results for selective and differential media demonstration.

Media Type	Organism Cultured	Reaction Observed	Selective Property	Differential Property

Questions

1. Define and give an example of a selective medium.

2. Define and give an example of a differential medium.

3. In some instances microbiologists use 'undefined growth media' such as nutrient agar to culture microorganisms and in other instances 'specialized media' are used. Indicate the circumstances for which each of these media types are appropriate.

Exercise 14: Urine Culture

Objective: To understand the laboratory methods used to diagnose urinary tract infections and to demonstrate the importance of a proper clean catch urine technique and transport of specimens. Upon completion of this exercise you should be able to:

- ❖ Define bacteriuria.
- ❖ Explain when bacteriuria is of concern.
- ❖ Explain how UTIs typically occur.
- ❖ Indicate which microbe is most often the cause of UTI.
- ❖ Describe the laboratory method used to diagnose a UTI.
- ❖ Calculate the number of microorganisms/ml present in a urine sample.
- ❖ List and explain 3 potential problems associated with collecting and transporting urine samples.
- ❖ Explain how to distinguish contamination of urine by normal flora from an infection.
- ❖ Explain when and why it is important to refrigerate urine samples.

Application: Diagnosis of urinary tract infections is dependent upon how the specimen is collected and processed. The parameters used for diagnosis dictate that the specimen be collected and transported properly.

Under normal circumstances the kidneys, bladder, ureters, urethra (with the exception of the opening) and the urine within are sterile. When excreted, urine can become contaminated by the normal flora of the opening of the urethra and external genitalia. Therefore, the presence of bacteria in voided urine (*bacteriuria*) does not necessarily indicate a urinary tract infection (UTI). Normal urine that is simply contaminated during passage out of the body contains very few bacteria (100-1,000 bacteria per milliliter is considered within the normal range) and often these bacteria are normal skin organisms that are not common pathogens of the urinary tract, thus easily identified as contaminants.

UTIs most often occur when normal intestinal flora present on the external urogenital surfaces become transplanted, invade the urethra, and then ascend into the bladder. In most cases these bacteria are opportunistic members of fecal flora. *E. coli* is by far the most frequent causative agent. In addition, urinary tract infections are the most common nosocomial infections accounting for approximately 39% of all cases. UTIs as nosocomial infections are usually associated with urinary catheterization.

Laboratory diagnosis of a urinary tract infection is made by culturing urine. To distinguish contamination of urine by normal urogenital flora from a urinary tract infection caused by the same organisms, it is necessary to determine the numbers of organisms present per milliliter of urine. In general, a count in excess of 100,000 organisms per milliliter of urine and the predominance of only one species in a properly collected and transported urine specimen is considered clinically significant. The presence of such large numbers of bacteria in urine correlates with active infection of the bladder or kidneys.

A quantitative culture is prepared by placing a measured volume of urine on an agar plate and counting the number of colonies that develop after incubation. A calibrated loop that delivers 0.01 ml (10µl) of sample is used to inoculate the plate. To determine the number of organisms per milliliter of urine, the colony count is divided by 0.01 ml (# colonies/0.1 ml). For example:

$$\textbf{21 colonies} \div \textbf{0.01 ml} = \textbf{2100 colonies / ml}$$

There are several problems associated with culturing urine samples. Because urine often becomes contaminated as it passes out of the body, culturing should begin soon after collection before contaminants can multiply and distort the results. Overgrowths in standing urine can lead to false positives. Since urine is an excellent medium for bacterial growth, samples should be immediately refrigerated. Bacterial counts in the refrigerator will remain constant for up to 24 hours. Also of concern is the potential for contamination by hardy organisms that can mask the presence of other pathogens that are difficult to cultivate on typical culture media. Either of these scenarios can lead to lab results that fail to accurately predict the clinical situation and possibly lead to the mismanagement of the patient's care. For these reasons it is important that urine specimens be collected using proper techniques and that the specimens be delivered to the laboratory for processing in a timely fashion.

Collection of urine for culture ---"clean-catch" technique:

1. Thoroughly clean all external surfaces surrounding the urethra.

2. Discard the first stream of urine. Collect the "midstream" portion in a sterile specimen container. The sterile specimen container should be held so it does not come into contact with skin or clothing.

3. Firmly place the lid on the container. Wipe the outside of the container with disinfectant.

4. Urine containers should never be filled to the brim. Double check and make sure the lid is on securely.

Procedure:

Lab 1:

14-A Urine Culture Techniques and Examination

1. Without any special care or technique, collect a specimen of your urine in the sterile container provided. Close the lid tightly and wipe the outside of the container with an alcohol wipe.

2. Use a sterile volumetric inoculating loop to collect 0.01ml of urine from your sample. Hold the loop vertically and immerse it in the urine sample. Carefully withdraw the loop to obtain the correct volume. The loop is designed to fill to capacity in the vertical position. Do not tilt it until you get the loop in position over the plate.

3. Inoculate a blood agar plate by making a single streak across the diameter of the agar.

4. Turn the plate 90° and with the inoculating loop spread the urine across the entire surface of the agar.

5. Turn the plate 90° and streak a third time to ensure that the specimen is distributed evenly across the plate.

6. Using the "clean catch" technique described in the exercise, aseptically collect a second sample of your urine. Wipe the outside of the container with an alcohol wipe and close the lid tightly.

7. Follow steps 2-5 to inoculate a second blood agar plate with the second urine sample.

8. Incubate the plates at 37° for 24 hours.

Lab 2:

14-A Urine Culture Techniques and Examination

1. Examine your plates for amount of growth, types of colonies, microscopic morphology, and Gram stain characteristics of colony types. Record your observations.

2. Count the number of colonies on each plate. Determine the number of organisms per milliliter present in each urine sample.

3. Record your results in **Table 14.1** and interpret any differences you observed in the amount of growth recovered from the two specimens.

14-B: Simulation of a False Positive Urine Specimen

1. A urine specimen was aseptically collected.

2. To simulate a typical false positive this specimen was inoculated with a relatively small number of *E. coli*. The sample was left out at room temperature for six hours.

3. At hourly intervals throughout the 6 hour incubation period samples were withdrawn and plated following the procedure outlined in **Exercise 14-A.**

4. The plates were incubated at 37° for 24 hours.

5. Count the number of colonies on each plate and record your results in **Table 14.2**.

6. Interpret the results obtained and include this information in the space provided on the **Results** page.

Results

Table 14.1 Record the results for Exercise 14A

	CFUs per milliliter	Number of colony types	Gram Reactions
Unclean urine			
Clean-catch, midstream urine			

14A: Interpret these results.

Table 14.2 Record the results for **Exercise 14B**, simulation of a false positive urine specimen.

	CFUs per milliliter
1 hr	
2 hrs	
3 hrs	
4 hrs	
5 hrs	
6 hrs	

14B: Interpret these results.

Questions

1. Define bacteriuria. When is bacteriuria significant?

2. What is one of the most common reasons urinary tract infections occur? Which organism is most frequently the causative agent?

3. What methods are used to diagnose a UTI?

4. Give detailed examples of two problems associated with collecting and transporting urine samples.

Exercise 15: Intestinal Pathogens

Objective: To become aware of the variety of intestinal pathogens, how they are manifested and identified. Upon completion of this exercise you should be able to:

- Explain why it is necessary to differentiate bacterial and protozoan infections of the GI tract.
- Describe how the media used in this exercise differentiate between bacterial pathogens and non-pathogens.
- Define enterotoxin.
- Differentiate between food infections & food intoxications.
- Define superinfection.
- Define trophozoite and cyst.

Application: Stool samples are collected and processed in different ways dependent upon the signs and symptoms the patient exhibits. For bacterial infections, stool samples are cultured and the pathogens distinguished by lactose fermentation characteristics. These pathogens are then further identified by biochemical means. Parasitic infections are diagnosed by preparing slides of preserved stool samples, and examining the slides for the presence of organisms. The identity of the pathogen is based on microscopic examination.

Many types of organisms (bacteria, viruses, yeasts, and parasites) can cause gastrointestinal upset. The majority are contracted by ingesting contaminated food or water. Bacterial agents often cause illnesses by producing enterotoxins (one type of exotoxin) either in the food prior to ingestion, or in the gut upon infection. *Staphylococcus aureus* is a common example of an organism that grows in food where it releases enterotoxin. The signs and symptoms of illness are entirely due to ingestion of the preformed toxin: *Staphylococcus aureus* does not grow in the gut. Shigellosis, on the other hand, is caused by an infection and growth in the gut by one of the four species of *Shigella*. However, most of the signs and symptoms of Shigellosis are caused by the production and release of enterotoxins in the gut by the infecting organism.

Viral infections of the gastrointestinal tract are probably the most common and range from diarrhea, nausea and vomiting, to hepatitis (inflammation of the liver). One of the most prolific is the Norwalk virus, which causes the illness popularly referred to as the "24-hour flu", even though the organism is not an influenza virus at all. Signs and symptoms include nausea, vomiting and diarrhea. These viral infections are extremely contagious but tend to last only a day or two.

Most gastrointestinal intoxications are self-limiting and treatment is based on the symptoms such as the dire need for water replacement in the case of cholera patients. Other afflictions are infections of the tissues of the intestine or colon and may require antibiotic intervention to clear the infection and avoid bacteremia/septicemia.

Diarrhea in some cases is caused by an overgrowth of yeast, commonly *Candida albicans*, in the intestine. This condition is most commonly due to a sudden decrease in the normal intestinal flora following treatment with a broad-spectrum antibiotic. An infection of this type is called a "superinfection" and usually resolves itself after antibiotic treatment ends and the normal gut flora reestablish.

Parasitic infections of the intestine usually cause diarrhea. With the more invasive organisms, this is often accompanied by cramping and bloody discharge. *Giardia lamblia* is a common intestinal parasite in the United States. Because it can be carried in the intestines of wild animals, *G. lamblia* is a common contaminant in remote water supplies. It is often contracted by hikers and backpackers who drink untreated water. However, in recent history, outbreaks of infections in daycare centers have been reported. This indicates *G. lamblia* can be passed from human to human by the fecal-oral route.

Intestinal parasites are identified by their morphological characteristics. For many of the protozoans two morphologically distinct stages, cysts and trophozoites, occur in the life cycle. The cyst is the dormant stage in the life cycle; the trophozoite is the actively growing or vegetative stage. The cyst and trophozoite stages for Entamoeba and Giardia are shown in **Figure 15.1.**

Figure 15.1 *Entamoeba histolytica*: intestinal parasite which causes diarrhea. Illness can progress to liver abscess.

(a) cyst   **(b) trophozoite**

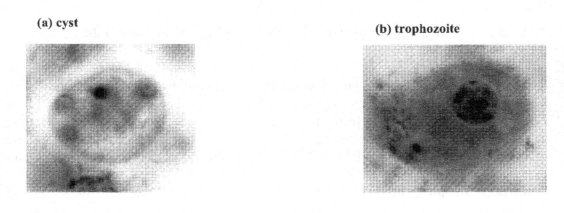

Figure 15.2 *Giardia lamblia* : intestinal parasite often acquired from remote water sources.

(a) cyst   **(b) trophozoite**

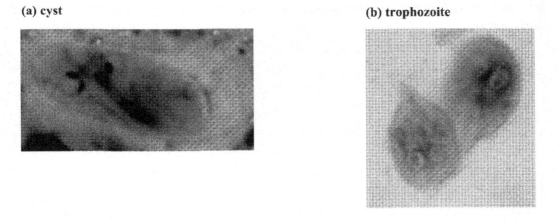

Procedure:

STOOL SPECIMEN

1. Bring a human stool sample (small- ~ the size of a pea) to class in the container provided.

2. The media you will be given select for Gram-negative bacteria and differentiate between lactose fermenters and non-lactose fermenters. These media are used for primary isolation of bacteria from stools because the majority of the bacterial pathogens that infect the gut are non-lactose fermenters. Using a sterile swab, inoculate the initial quadrant of each plate of media you have been given and streak the plates for isolation.

3. Incubate the plates at $37^{\circ}C$ for 18-24 hrs.

INTESTINAL PARASITES

1. Several prepared slides of intestinal parasites will be available for viewing. Using the information and pictures provided, identify each of the pathogens and their stage of presentation.

Results

Stool Specimen

1. Are there any non-lactose fermenters in your stool cultures?

2. If there were non-lactose fermenters present, based on your signs and symptoms, would you expect them to be intestinal pathogens?

3. What would they be if not intestinal pathogens?

Intestinal Parasites

1. Draw each of the parasitic organisms seen in the prepared slides. Take care to add the features that allowed you to identify each pathogen.

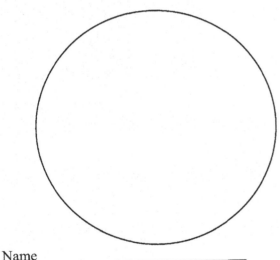 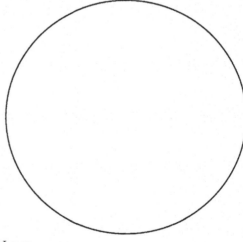

Name _____

Flagellate or Amoebae?

Illness _____

Slide #_____

Name _____

Flagellate or Amoebae?

Illness _____

Slide #_____

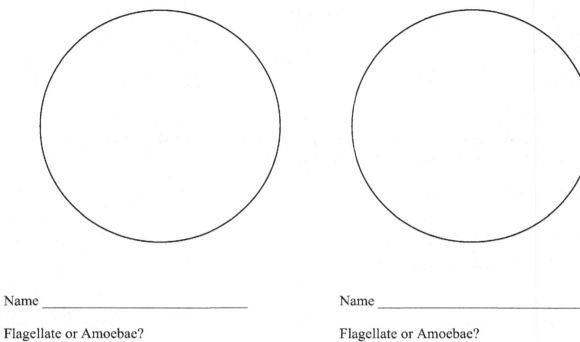

Name _____

Flagellate or Amoebae?

Illness _____

Slide #_____

Name _____

Flagellate or Amoebae?

Illness _____

Slide #_____

Questions

1. Describe how bacterial and parasitic infections are diagnosed from stool specimens.

2. Describe the distinguishing metabolic properties of most bacterial pathogens.

3. Define trophozoite and cyst.

Exercise 16: Staphylococcus

Objective: To study isolation and identification techniques for *Staphylococcus* sp. Upon completion of this exercise you should be able to:

- ❖ Identify where *Staphylococcus* is part of the normal flora.
- ❖ Explain why staphylococcal infections are significant.
- ❖ Differentiate between *S. aureus* and *S. epidermidis* in terms of normal location and pathogenicity.
- ❖ Describe how MSA is selective and differential.
- ❖ Explain what tests are used to determine if a given *Staphylococcus* isolate is pathogenic.
- ❖ Explain why the production of coagulase is an advantage for pathogenic *Staphylococcus* species.

Application: *Staphylococcus aureus* infections are very common and potentially very severe. It is a common cause of nosocomial infections and hospital strains are often noted for their extreme antibiotic resistance. In recent years, some of the resistant strains that were once limited to hospitals have become common isolates from community-acquired infections. Awareness of the potential of this pathogen is essential for anyone associated with the medical field.

The genus *Staphylococcus* is composed of Gram-positive cocci characteristically arranged in irregular "grape-like" clusters. They are ubiquitous in the environment and an important part of our body's normal flora on skin, mucosal surfaces, and in the upper respiratory tract.

The genus *Staphylococcus* contains some of the most important disease causing organisms in humans. They are members of a group of Gram-positive bacteria called the pyogenic (pus-producing) cocci. Infections commonly occur when staphylococcal strains become transplanted to a normally sterile part of the body as a result of trauma or abrasion to the skin or mucosal surface. Staphylococci are easily transmitted from person to person. Upon transmission the bacteria may become established as part of the recipient's normal flora. Person to person spread of staphylococci, especially antibiotic resistant strains, is of great concern in hospitals and presents substantial infection control problems. Nosocomial infections caused by contamination with staphylococci transmitted during surgery and other invasive procedures are of great concern.

Two common, clinically important species are *Staphylococcus epidermidis* and *Staphylococcus aureus*. *S. epidermidis* is most often found as a member of the normal flora residing on the skin and mucous membranes. It is not usually pathogenic, but may cause opportunistic infections. Most staphylococcal disease is caused by strains of *S. aureus*. At least 20% of healthy individuals carry *S. aureus* as part of their normal flora, usually in their nostrils.

Laboratory diagnosis of staphylococcal disease is made by identifying the organism in a clinical specimen taken from the infection site (pus from skin lesions, sputum when pneumonia is suspected, urine, spinal fluid, or blood). Clinical specimens may contain mixed normal flora as well as the pathogenic bacteria responsible for the infection. When cultivated together, Gram-negative organisms often inhibit the growth of the Gram-positive organisms present in the sample. Therefore, it is necessary to use selective media designed to inhibit Gram-negative organisms when attempting to isolate Gram-positive pathogens from mixed cultures.

Mannitol Salt Agar (MSA) is a selective and differential medium. It is selective for staphylococci and can presumptively differentiate pathogenic staphylococci (*S. aureus*) from other staphylococcal species (Note: Some strains of *Staphylococcus aureus* are mannitol negative, therefore in the clinical setting it is necessary to perform the definitive test for coagulase). MSA contains mannitol (a sugar), 7.5% NaCl, and phenol red (a pH indicator). The NaCl concentration makes the media selective for staphylococci, since most other bacteria cannot survive the high salinity. MSA differentiates staphylococcal species based on their ability to ferment mannitol. Most strains of *S. aureus* (pathogenic) ferment mannitol. This results in acid production which turns the pH indicator from red to yellow. Nonpathogenic staphylococci (*S. epidermidis*) are able to grow on MSA but do not ferment mannitol and therefore the media remains red.

The definitive test for pathogenic staphylococci is detection of the coagulase enzyme. *S. aureus* is an opportunistic pathogen that can be highly resistant to normal immune responses and antibiotics. This resistance is in part due to the production of coagulase. Coagulase is an enzyme which works with components in serum to form protective fibrin barriers (clots) around the bacterial cells to shield them from the body's immune responses. The coagulase test is performed by adding the test organism to rabbit plasma. If the organism is positive for coagulase, the plasma will thicken or develop protein clots within 4 hrs at 37°C.

Procedure:

Lab 1
Identification of Staphylococci

1. Obtain a Mannitol Salt Agar plate. Divide the plate into 2 sections.

2. Inoculate one side of the MSA plate with *S. epidermidis* and the other with *S. aureus*.

3. Incubate at 37° for 24 hours.

4. Divide a blood agar plate into 2 sections.

5. Inoculate one side of the BAP with *S. epidermidis* and the other with *S. aureus*.

6. Obtain 2 tubes containing rabbit plasma. Inoculate one tube with S. *epidermidis* and the other with *S. aureus*.

Lab 2
Identification of Staphylococci

1. Examine the MSA & BAP plates. Note the colony types and mannitol reactions for each.

2. Indicate the Gram stain reaction and microscopic morphology of each colony type.

3. Record your results in **Table 16.1**.

Results

Table 16.1 Identification of Staphylococci species results.

	Colony morphology	Gram stain	Mannitol	Coagulase
Staphylococcus aureus				
Staphylococcus epidermidis				

144

Questions

1. Why are *Staphylococcus aureus* infections of such great concern, especially in the hospital setting?

2. What are the potential problems involved in isolating Gram-positive organisms from patient specimens?

3. What properties make MSA medium selective?

4. What properties make MSA medium differential?

5. What is the definitive test for diagnosing pathogenic staphylococci?

6. How does the production of coagulase by *S. aureus* increase virulence?

Exercise 17: Streptococcus

Objective: To become familiar with the pathogens of the respiratory tract and learn how to distinguish them from the normal flora. Upon completion of this exercise you should be able to:

- ❖ Identify where on the body *Streptococcus* occurs as normal flora.
- ❖ Differentiate between *S. pneumoniae* and *S. pyogenes* on the basis of transmission and pathogenicity.
- ❖ Differentiate between   and   hemolysis.
- ❖ List and explain the test used to differentiate *Staphylococcus* from *Streptococcus*.
- ❖ Describe what tests are used to differentiate pathogenic   hemolytic *S. pyogenes* from other   hemolytic *Streptococcus* species.
- ❖ Describe what tests are used to differentiate pathogenic   hemolytic *S. pneumoniae* from other   hemolytic *Streptococcus* species.

Application: Normal flora of the nose and throat includes many species of the genus *Streptococcus*. However, *Streptococcus pyogenes* is one of the primary pathogens of the throat. Thus an important job of the clinical laboratory scientist involves distinguishing pathogenic organisms from components of the body's normal flora.

The genus *Streptococcus* is composed of Gram-positive cocci characteristically arranged in chains. Many streptococcal species are found as part of the normal flora of the human body, predominantly in the upper respiratory tract.

The throat is home to many types of bacteria. The most common bacterial species in the throat are nonpathogenic viridans streptococci, coryneforms, *Staphylococcus* sp., and the opportunistic pathogen *Streptococcus pneumonia*. *S. pneumoniae* is found in the throat of a large percentage of healthy people. The most common scenario of this opportunist is that *S. pneumoniae* enters the lung as a secondary infection to cause pneumonia when host defenses have been weakened by a primary infection, caused by some other agent, such as an influenza virus. *Streptococcus pyogenes*, the organism which causes "strep throat" is a true pathogen (rather than an opportunist) of the throat. It is commonly transmitted via aerosol droplets from infected individuals.

Clinically, the identity of the majority of the normal flora of the respiratory tract is unimportant. The most significant duty of the clinical microbiologist in this case is distinguishing the pathogenic species from the normal flora. The streptococci are initially distinguished on the basis of their hemolytic activity. When grown on blood agar, one of three patterns of hemolysis occurs:

1) -hemolysis: hemolysins cause complete lysis of red blood cells. The result is a clear zone around colonies.
2) -hemolysis: hemolysins cause partial lysis of red blood cells. The result is a green zone around colonies.

3)   -hemolysis: no hemolysis. This results in no change in the blood agar.
-hemolytic streptococci are further differentiated based on their susceptibility to the antibiotic bacitracin. Bacitracin-susceptible, -hemolytic streptococci are pathogenic and are referred to as Group A streptococci, or *Streptococcus pyogenes*. Over 90% of streptococcal infections are caused by *S. pyogenes*.

-hemolytic streptococci are further differentiated based on their susceptibility to optochin. Optochin-susceptible,   -hemolytic streptococci are *S. pneumoniae*. Optochin-resistant,   - hemolytic streptococci are considered nonpathogenic viridans streptococci.

Catalase is an enzyme produced by *Staphylococcus* sp. but not by *Streptococcus* sp. and therefore can be used to distinguish between streptococci and staphylococci. The enzyme catalyzes the break down of hydrogen peroxide to water and oxygen, which is liberated from the reaction.

Procedure:

Lab 1
Identification of Streptococci

1. For these tests each student will be assigned either a   or   -*Streptococcus* species and a control organism to analyze.

2. Inoculate the assigned *Streptococcus* species onto on a blood agar plate (BAP) and streak for isolated colonies.

3. Place the appropriate disk (bacitracin or optochin) within the first quadrant of the streak plate.

4. Label the plate.

4. Incubate the plates at 37°C for 24 hours.

5. Repeat steps 2-4 above for the assigned control organism.

6. Perform the catalase test on a *Staphylococcus* species and your assigned *Streptococcus* species according to the following procedure:

   a. Place a drop of hydrogen peroxide on a glass slide.
   b. Place a small amount of bacterial culture in the hydrogen peroxide drop. Be careful not to pick up any blood agar along with the bacterial cells. Since blood cells contain catalase, addition of blood to hydrogen peroxide will give a false positive result.
   c. The appearance of bubbles is a positive test result for catalase.

7. Record the results in **Table 17.2**.

Lab 2
Identification of Streptococci

1. Examine the BAP for hemolysis.

2. Are your assigned bacteria resistant, or susceptible, to the antibiotic (optochin or bacitracin) used? Based on their resistance, which species do you think you have?

3. Indicate the Gram stain reaction and microscopic morphology of the specimen

4. Record the results in **Table 17.1**.

Results

Table 17.1 Identification of Streptococci.

	Describe hemolysis	Disk used/reaction	Gram Reaction
α-streptococci			
α-streptococci control			
β-streptococci			
β-streptococci control			

Table 17.2 Catalase test results.

	Catalase Reaction
Streptococci	
Staphylococci	

Questions

1. Where on the body would you expect to find *Streptococcus* species as part of the normal flora?

2. What biochemical test is used to distinguish *Staphylococcus* species from *Streptococcus* species? Describe the reaction and the product of the reaction.

3. Why is it important when taking a throat culture to avoid touching the swab to any tissue other than that of your throat? How could incidental contact with your lips or the side of your mouth affect the results obtained?

4. List and describe the three patterns of hemolysis discussed in the exercise. Give one example of a bacterium that produces each of the three patterns.

Exercise 18: Nasal and Throat Cultures

Objective: To perform nasal and throat cultures. To observe typical results of nasal and throat cultures.

Application: Throat cultures are among the most common aerobic cultures ordered by physicians. Nasal cultures are not as common but can lend important information in certain epidemics.

A throat culture is performed on a patient to determine if *Streptococcus pyogenes* is present. This organism is medically important because the body's immune response to a *S. pyogenes* infection may result in cross reactions that damage body cells. Damage to heart cells may result in rheumatic fever or in the case of damage to kidney cells, glomerulonephritis. Isolation of this organism in any quantity is considered pathogenic and indicates the patient should be treated. Occasionally *Streptococcus pyogenes* does occur as an asymptomatic colonization of the throat; but even a patient who is an asymptomatic carrier should be treated to prevent transmission of the disease

While normal flora in the nose mirrors what is seen in the throat quite well, *Streptococcus pyogenes* is rarely found in the nose. For this reason nasal cultures are not as common as throat cultures. However, about 20-25% of the general population does carry *Staphylococcus aureus* in the nose. In recent years, methicillin-resistant strains of *Staphylococcus aureus,* designated MRSA, have been detected. Since about 2001, MRSA has become more and more common and occurs asymptomatically in the nose in 1-2% of the population. Under the right circumstances MRSA can be a serious pathogen. Some strains appear to have acquired virulence mechanisms that are more advanced than those of the less complicated *Staphylococcus aureus.* MRSA strains are often responsible for skin infection epidemics so when a patient is suspected of carrying methicillin-resistant *Staphylococcus aureus* a nasal sample is taken. Populations at greater risk for MRSA infections are those that play contact sports and those in closed societies such as prisons and the military.

MRSA diagnosis will be performed using MSA medium to avoid isolation of the normal flora of the throat. The identification of pathogenic strains is based on resistance to methicillin, oxacillin or cefoxitin. The antibiotic resistance is used as an indicator of other, as yet unknown, virulence mechanisms unique to MRSA and generally not found in average *Staphylococcus aureus* strains.

Many organisms live in the throat as normal flora. The colony morphology of common normal flora that you are most likely to see growing on a blood agar plate is described below.

MOST ABUNDANT

α-hemolytic streptococci (alpha): appear as very small translucent to white colonies with a green zone of hemolysis. This group may include *Streptococcus pneumoniae* which is normal in the throat. Colony variations for *S. pneumoniae* may include heavy capsular material, thus a mucoid colony type, or a colony which is sunk in at the center (checker-shaped).

streptococci (gamma): appear as very small translucent to white colonies with no zone of hemolysis

MODERATELY ABUNDANT

Neisseria meningitidis, other *Neisseria* sp, or *Moraxella catarrhalis* appear as grey to beige/brown colonies. In Gram stain preparations the organisms appear as Gram negative diplococci.

Other *Moraxella* species appear as small beige to brown colonies but are Gram negative coccobacilli (short plump rods).

The coryneforms comprise a group of bacteria which form small white colonies which often appear to "burrow" into the agar. In a Gram stain they appear as Gram positive club shaped (pleomorphic) rods. Cells clump and line up side by side (a palisade arrangement) and have been described as looking like Chinese alphabet characters.

Staphylococcus sp appear as medium sized opaque white colonies with or without β-hemolysis.

RARELY ABUNDANT

Enteric bacteria appear as large creamy white to slightly yellow colonies although colony morphology can vary greatly. In Gram stain they appear as Gram negative rods.

Yeasts appear as small opaque white colonies. Yeast colonies can usually be determined by other colony characteristics. The surface of the colony has a ground glass appearance and rarely, the colony has an irregular edge indicating the formation of pseudohyphae.

NOTE: *Staphylococcus aureus* and many of the Gram-negative enteric bacteria produce β-hemolytic zones and must be differentiated from *Streptococcus pyogenes.*

Procedure:

Throat Culture
Day 1

1. Take a culture by swabbing the back of your throat with a sterile swab. Avoid touching any other part of your oral cavity.

2. Inoculate one quadrant of a BAP and streak for isolation. Dispose of the swab in the biohazard bag provided.

3. Incubate the BAP in a candle jar* at 37°C for 24 hours.

Throat Culture
Day 2

1. Examine your throat culture plate. Pick 2 different isolated colony types for further characterization.

2. Inoculate a blood agar plate with the first colony. Remember to streak the plate for isolated colonies.

3. Inoculate a blood agar plate with the second colony. Remember to streak the plate for isolated colonies.

4. Utilizing the Flow Chart (**Figure 18.1**), characterize and, if possible, identify each of your throat isolates.

5. Record your results in **Table 18.1**.

*Many bacteria that colonize the human body grow best if placed in an increased CO_2 environment. An effective method for creating this environment is to seal the media in a jar with a burning candle. When the candle goes out the oxygen level is reduced from 21% (atmospheric) to 17%. The CO_2 level is increased to between 5-10%.

Nasal Swab
Day 1

1. Take a culture by swabbing the inside of one nostril with a sterile swab.

2. Inoculate the nasal swab in the top quarter of an MSA plate. Streak the plate for isolated colonies.

3. Incubate the MSA plate at 37°C for 24 hours.

Nasal Swab
Day 2

1. Examine your nasal culture plate and pick 2 different isolated colony types for further characterization.

2. For each isolate note the colony morphology and mannitol reaction.

3. Perform a Gram stain on each isolate.

4. Perform a coagulase test on any presumptive *S. aureus* colonies. Note: If your sample is coagulase positive see your laboratory instructor for further information and instruction.

5. Report your results in **Table 18.1**.

Figure 18.1: Flow chart outlining the sequence of basic tests necessary to characterize and identify a throat culture specimen.

Original Throat Culture
Blood Agar Plate

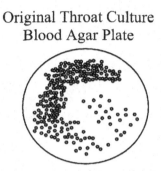

1. Choose 2 isolated colonies to characterize. Subculture each isolate to a fresh blood agar plate. Streak for isolation.

Note: if the isolate you choose is hemolytic place the appropriate antibiotic disk (Bacitracin or Optochin) in the first quadrant of the streak plate.

Throat Isolate #1   Throat Isolate #2

2. Incubate plates in a candle jar at 37°C for 24-36 hours.

3. For each isolate determine: colony morphology, Gram stain reaction, and cellular morphology.

Gram positive cocci

Characterize based on information contained in this lab exercise.

Catalase Test

+ / -

Coagulase Test   *Streptococcus species*

+ / -

Staphylococcus aureus   *Staphylococcus species*

Results

Table 18.1 Results of tests on isolates from throat and nasal swabs.

	Colony morphology	Gram reaction	Catalase	Hemolysis	Mannitol	Coagulase
Throat Isolate #1						
Throat Isolate #2						
Nasal Isolate #1						
Nasal Isolate #2						

Use the results listed above and the general information in **Exercises 16, 17,** and **18** to name (to species if possible) each of the four isolates obtained from your throat and nasal cultures. If you cannot name the genus and species then give a viable description of the organism's characteristics. Explain your strategy for characterizing your isolates.

Exercise 19: Wounds

Objective: To learn the proper technique for disinfection of a wound and for specimen collection. Upon completion of this exercise you should be able to:

- ❖ Define and differentiate bacteremia and septicemia.
- ❖ Differentiate between local, focal, and systemic infections.
- ❖ Explain how a bacterial infection of a wound could cause a systemic illness without infecting the bloodstream and give an example of this process.
- ❖ Explain why anaerobes, aerobes, and skin microbes might all be present in a wound.
- ❖ Explain why it is important to disinfect a wound before culturing the wound.
- ❖ Explain why wounds are disinfected using a spiral pattern.

Application: Wounds can become infected with skin flora and occasionally with multiple organisms. It is important for therapy to be able to interpret whether the organisms isolated from a wound are the infecting organisms or skin contaminants.

Wounds become infected with many types of microorganisms: aerobes, anaerobes, Gram positive, Gram negative, and sporeformers. The size of the wound may play a role in the likelihood of infection but is not a significant parameter in determining the etiological agent or the severity of the infection. Some bacterial infections do not need to be large to spread through the tissue, into the bloodstream, meninges, or to cause focal infections. Spread of infection is dependent on the virulence mechanisms of the microbe and the resistance factors of the host.

In some wound infections, the organism never leaves the site of the wound and yet a systemic illness leading to a fever and other severe effects can result. Usually this scenario is due to the production and release of exotoxins by the bacterium. In most cases, exotoxins have very specific modes of action and thus require only very low concentrations to cause severe results. Tetanus toxin is a good example of this. When *Clostridium tetani* infects a wound, the organism never leaves the site of the wound and the neurotoxin it releases causes all of the signs and symptoms of the disease. The neurotoxin is protein and soluble in body fluids so it can travel throughout the body in the blood and lymphatic systems. The neurotoxin binds to nerve synapses and initiates nerve impulses. As a result, muscles contract and cannot relax. The patient is unable to breathe and without medical intervention death may occur.

Deep wounds create the perfect environment for anaerobic bacteria. Many infectious anaerobic agents can be successfully cultured but the specimens should be transported to the laboratory in special anaerobic transport medium. A few anaerobic infections are caused by unculturable organisms. For these organisms, atmospheric oxygen is toxic and they cannot survive even the brief exposure to oxygen during the time required to move a swab or tissue specimen from the wound into an anaerobic transport medium.

Before collecting a wound specimen, the skin around the wound must be disinfected. This is a very important step because wounds can become infected with skin flora and it is important to be able to distinguish between organisms causing infection and normal flora. The disinfection process is performed by washing the wound and surrounding area with an antiseptic. When washing the area use an outward spiral motion beginning at the center of the wound. With this technique the most concentrated disinfectant is deposited near the wound site and as the swab is moved outward, encountering skin flora, the skin flora is moved away from the wound.

Procedure:

1. You will be issued a recently infected chicken leg.

2. Disinfect the surface of the wound making a spiral motion outward from the wound with a Povidone-Iodine swabstick. Repeat the disinfection.

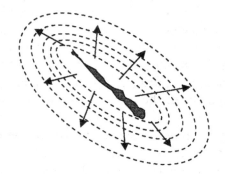

3. Use a dry sterile swab to open the wound and obtain a sample from its interior surface. Plate the sample on an agar plate and steak for isolation.

Results

1. Describe the organisms isolated. Give colony characteristics and Gram stain results.

2. Interpretation: Which organism do you think is the infecting organism and why?

166

Questions

1. What is the purpose of the pattern used to disinfect around a wound?

2. What term designates an infection which spreads from a wound to the blood stream?

3. How might a bacterial infection cause a systemic illness without leaving the wound?

Exercise 20: Dental Microbiology

Objective: Learn how dental caries are formed and from your own saliva isolate the organisms responsible. Upon completion of this exercise you should be able to:

- ❖ Describe how *Streptococcus* species cause dental caries.
- ❖ Explain why *S. mutans* is incubated in a candle jar.
- ❖ Explain what properties make MSB selective.
- ❖ Describe the colonies of *S. mutans* on MSB.
- ❖ Describe the colonies of *S. mutans*, *S. salivarius,* and *S. sanguis* on SBAP.
- ❖ Calculate the number of colonies per ml saliva.

Application: Three species of streptococci have been well studied for their role in the production of dental caries. These organisms are abundant in the saliva of every individual.

Living within the oral cavity is a population of microorganisms which constitute the normal flora. Everyday the oral cavity is exposed to many types of bacteria from external sources such as soil, food, and human skin. Most likely by means of competition, the normal flora prevents other organisms, such as those found in the foods we eat, from colonizing and infecting the oral cavity. Interestingly, normal oral flora includes many members of the genus *Streptococcus*. The roles of *Streptococcus mutans*, *Streptococcus sanguis*, and *Streptococcus salivarius* in the formation of dental caries have been extensively studied.

The species of *Streptococcus* that inhabit the oral cavity are capable of using sucrose (table sugar) as their carbon and energy source by the process called fermentation. The primary byproduct of this fermentation reaction is organic acid. This acid is capable of causing serious damage to the enamel of the teeth and thus can cause cavities (dental caries). However, there is more to the story than the production of organic acid alone. If it was not for the presence of a biofilm produced by the bacteria, saliva would simply wash the organic acid from the surface of the tooth and no tooth decay would occur. As part of their normal metabolism, the *Streptococcus* sp. in the oral cavity form biofilms on the teeth. An important component of this biofilm is dextran, a very large polymer of glucose. Dextran is manufactured by the bacteria from the table sugar (sucrose) in the sweets we eat. Unlike simple sugars which are water soluble, dextran has the properties and consistency of sticky gum. It accumulates on the teeth as the material known as plaque. The streptococci reside in, and under the plaque, and continue to use the dextran as a food source. The result is production of organic acids. If these acids remain in contact with the tooth surface for long periods of time, the enamel dissolves and dental caries are produced. Effective brushing and flossing of teeth removes plaque.

Biofilms are not unique to teeth. They commonly form on surfaces that are permanently submerged in water, such as, the hull of a ship or the bottom of a fountain. A good example of a biofilm is the slimy layer found on rocks in a stream. The slimy coat is due to microbial growth and consists of materials such as polysaccharides and polypeptides that are produced and excreted by the organisms. In all likelihood, these cell products insure that the microbe remains stuck to the rock as the water streams past. A medically important biofilm is one which forms on the surface of catheters that remain in place for long periods of time. These biofilms can harbor organisms which cause urinary or systemic infections in catheterized patients.

In this exercise you will isolate *S. sanguis* and *S. mutans* from your oral cavity. When these bacteria are grown on the surface of agar, the slimy substances they produce results in a very mucoid colony morphology, aiding in their identification.

Procedure:

20A *Streptococcus mutans* Isolation

1. Chew a piece of chewing gum until the flavor is gone. Don't swallow. The purpose is to collect saliva for the isolation of normal oral flora.

2. Roll a sterile tongue depressor around in your oral cavity until both sides are thoroughly wet with saliva. Draw the tongue depressor out of your mouth through closed lips to remove excess saliva but do not swallow yet. (Refer to **20B** for further instruction). Lay one side of the depressor on the agar of a MSB plate. Then turn the depressor over on the surface of the plate to inoculate with the other side of the blade.

   MSB is mitis salivarius bacitracin agar which inhibits the growth of most oral bacteria except *S. mutans*. It also contains sucrose to promote capsule formation.

3. Incubate the plate in a candle jar* for 72-96 hrs at 37°C.

4. Because there are a few other bacteria that can grow on MSB colony morphology is used to distinguish *S. mutans*. Observe for small, light blue to black, raised and rough colonies. The colonies resemble etched glass or burnt sugar.

*Many bacteria that colonize the human body grow best if placed in an increased CO_2 environment. An effective method for creating this environment is to seal the media in a jar with a burning candle. When the candle goes out the oxygen level is reduced from 21% (atmospheric) to 17%. The CO_2 level is increased to between 5-10%.

20B *Streptococcus salivarius, S. mutans,* and *S. sanguis* Isolation

1. Collect your saliva in the cup provided.

2. Use a 10 µl calibrated loop to sample your saliva and make a single streak down the surface of a sucrose blood agar plate. Then streak perpendicular to the initial inoculation line to cover the entire surface of the plate.

3. Incubate the plate in a candle jar for 72-96 hrs at 37°C.

4. On this medium, *S. salivarius* produces large, raised, mucoid colonies. *S. sanguis* produces glistening colonies surrounded by an indentation of the agar surface. These colonies cannot be moved without tearing the agar. *S. mutans* produces colonies that are recognized by the presence of a drop of liquid polysaccharide on top of or surrounding the colony.

5. Count the number of each colony type and determine the number of colonies per ml of saliva.

cfu/ml = number of colonies / amount plated

Results

1. Describe the colonies observed on the MSB plate. Can you identify colonies of *S. mutans*?

2. Describe the colonies observed on the sucrose blood agar plate.

3. For each colony type observed on your sucrose blood agar plate, calculate the number of colonies per ml of saliva.

Questions

1. Describe the process of caries formation.

2. Describe the appearance of *S. mutans* on MSB.

3. Describe the appearance of *S. mutans*, *S. sanguis*, and *S. salivarius* on sucrose blood agar.

Exercise 21: *Neisseria gonorrhoeae*

Objective: To become familiar with the diseases and laboratory diagnosis of *Neisseria gonorrhoeae*. Upon completion of this exercise you should be able to:

> ❖ List 3 characteristics of *Neisseria gonorrhoeae*.
> ❖ Explain why you cannot get gonorrhea from toilet seats.
> ❖ Define reportable disease.
> ❖ Describe how *N. gonorrhoeae* is identified.
> ❖ Describe how *N. gonorrhoeae* could be presumptively identified at the bedside.

Application: *Neisseria gonorrhoeae* is one of the most common sexually transmitted diseases.

Neisseria gonorrhoeae is a fastidious, Gram-negative diplococcus that is an obligate human pathogen. This means that it does not live outside the human body and to survive it must be transmitted from one human to another. It is one of the most common sexually transmitted diseases, although the genitals are not the only area that can become infected. *N. gonorrhoeae* can infect any of the mucus membranes, including the conjunctiva of the eyes and transmission by contact with fluid from such infected sites can occur.

Due to the ease of transmission from one human to another, *N. gonorrhoeae* is one of the organisms on the list of reportable diseases. By federal law, diagnosis of a disease on that listing must be reported to state and/or federal agencies. They are therefore called "reportable diseases". As illustrated by **Figure 21.1** *Neisseria gonorrhoeae* occurs as an intracellular diplococcus and detection of diplococci within the polymorphonuclear leukocytes (PMNs) is a diagnostic characteristic.

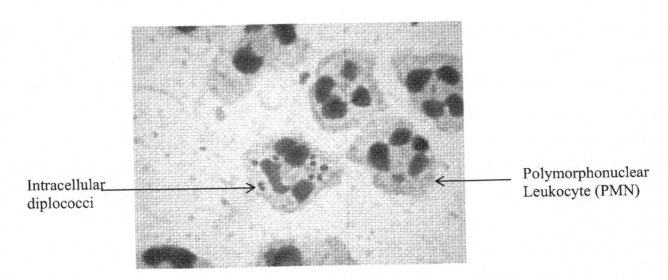

Figure 21.1 Microscopic field showing intracellular diplococci within a polymorphonuclear cell.

Procedure:

N. gonorrhoeae is normally plated on special medium with antibiotics to select for its growth. It is then identified on the basis of the biochemical reactions of the culture. However, at the bedside, *N. gonorrhoeae* can be presumptively diagnosed from a Gram-stained patient sample if Gram-negative diplococci are seen within polymorphonuclear leukocytes (PMNs). The presence of these diplococci within host cells indicates *N. gonorrhoeae* is provoking an immune response and has been phagocytized by the PMNs.

Prepared Slides

Prepared slides from patient specimens are available for viewing as a demonstration.

Results

1. Draw your observations in the circle below: Be sure to show the relationship to the host cell.

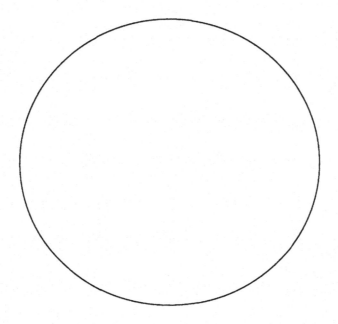

Questions

1. List 3 characteristics (morphological and cultural) of *Neisseria gonorrhoeae*.

2. Define a "reportable disease".

3. Under what circumstances can *Neisseria gonorrhoeae* be diagnosed by a Gram stain?

Section IV: Public Health and Environmental Microbiology

The health of the general public depends upon the maintenance of safe food and water supplies. Public health facilities in the United States are the most comprehensive in the world. Our water is treated and tested to ensure that there is safe drinking water in every home. Sewage disposal is controlled and laws prohibit the dumping of sewage or chemical pollutants into all of our waterways. The intent is to protect our environment and make a safer place to live.

Laws aimed at protecting the consumer from contamination also regulate the composition, handling, and distribution of food. Food manufacturing industries are under mandatory laws for testing of their products to ensure quality and safety.

Safeguarding fresh foods, such as meats, is more difficult to control since contamination by material from the intestine of the animal is a major concern. In many cases we rely on proper cooking to ensure the safety of the consumer. This is an imperfect means of control and improper cooking is the most likely cause in most cases of food poisoning. However, state governments have cooking regulations in place for restaurants and state inspectors help ensure that these regulations are followed.

The safety of milk and milk products is normally ensured by pasteurization. This involves heating the milk to moderately high temperatures, a treatment which reduces the total number organisms in the milk without significantly changing its quality. By reducing the total number of microorganisms in the milk, the numbers of potential pathogens are also significantly reduced or eliminated.

This section of the manual will introduce you to standard methods for examining food and water. The purpose is to understand how these methods ensure the quality of the final products we consume.

The study of environmental microbiology is quickly becoming one of the 21st Century's primary scientific study areas. The agriculturally significant microbes of past studies have been somewhat limited to nitrogen fixers because of their importance in agriculture. However, as we enhance our understanding of symbiotic relationships, we have realized that the relationships between certain microbes, and between plants and microorganisms, are equally important to understanding how to improve agricultural yields. Relationships of microbes to changes in the earth's surface are just being realized as we work to better understand the major groups of geomicroorganisms. The incredibly diverse collection of environmental microbes promises to greatly improve our understanding of geological formations, soil formation, and agricultural techniques as they are identified and their relationships to each other and the earth are further characterized.

Exercise 22: Food Microbiology

Objective: To understand the factors responsible for food poisoning outbreaks, to know the characteristics of organisms that cause these outbreaks, and the various effects they have on humans. Upon completion of this exercise you should be able to:

> ❖ Give 2 reasons why food manufacturers decontaminate food as a routine part of processing.
> ❖ Define and differentiate food infections and food intoxications and give an example of each.
> ❖ Explain why it is safe to eat a rare steak but not a rare hamburger.
> ❖ Explain why different media are used to isolate bacteria from different foods.

Application: Understanding the pathogenic character of the microorganisms that cause food poisoning helps in the diagnosis, treatment, and prevention of these illnesses.

Most foods naturally have normal microbial flora. Foods such as vegetables and fruits have microbes from the soil on their outer surfaces, but they are sterile on the inside. Most organisms that contaminate produce are harmless when ingested by humans but play a role in spoilage of the product. However, the normal flora of fruits and vegetables often becomes important in processing and preserving foods. In some cases, if these microbes are allowed to grow to high numbers, they produce substances in the food product that are harmful when ingested. Consequently, many manufacturers decontaminate the food substances as part of processing to reduce the risk to the consumer and to increase the shelf life of their product.

Meats and poultry would be sterile if it were possible to use aseptic technique during slaughter. But the butchering process is very messy and often the contents of the animals gut are allowed to contaminate the meat. In most cases this is not a problem because meat is normally cooked before consumption and this kills the contaminating microbes. However, with ground meats the problem is exacerbated because fecal contamination on the surface of the meat gets mixed throughout the product. Ground meat is then only truly safe to consume after it has been cooked thoroughly and the interior temperature has reached 60°C to ensure that the contaminating microbes are dead.

Cases of food poisoning are not uncommon in the United States but usually only one or a few people are affected. However, the potential for large outbreaks (epidemics) exists especially since restaurant chains typically purchase large quantities of food from a single source and distribute it widely. For example, in 1993 in the Northwestern states, Washington, Oregon and Idaho, there was a significant outbreak of a very toxic strain of pathogenic *E. coli*. Many people were hospitalized and several died as a result of eating contaminated hamburger.

There are two types of food poisoning: food infections and food intoxications. Food infections are caused by organisms that enter and multiply in the intestine. The resulting infection may cause fever in addition to gastrointestinal upset. Pathogenic *E. coli*, *Salmonella*, *Shigella* and *Campylobacter* are common genera that cause food infections. Food intoxications, on the other hand, result from the ingestion of preformed toxins. Such toxins are produced when bacteria grow to high numbers in the food. To become ill, it is not necessary to ingest the organism at all, only the toxin. *Staphylococcus aureus* and *Clostridium botulinum* are microbes that cause food intoxications.

Procedure

1. Solid foods such as meats, and cheeses
   a. Weigh out 10 grams of the product being tested.
   b. Place this sample in a blender.
   c. Add 90 ml of sterile water & blend.
   d. Remove 0.1 ml of the suspension and plate.
   e. With a sterile hockey stick spread the suspension over the entire surface of the plate.

2. Dried foods such as grains, and spices
   a. Weigh out 10 grams of the product being tested.
   b. Add this sample to a dilution bottle containing 90 ml sterile water.
   c. Shake vigorously 25 times.
   d. Remove 0.1 ml and plate.
   e. With a sterile hockey stick spread the sample on plate.

3. Fresh or dried fruit
   a. Dampen a sterile cotton swab in sterile water.
   b. Swab the outside of the product.
   c. Inoculate a plate of media and steak the entire plate to obtain plate counts.

4. After incubating your plates for 48 hours, at a temperature determined by your instructor, examine the plates for growth of organisms.

5. Count the colonies on plates that have between 30 and 300 colonies

6. Calculate the total number of organisms per gram of food or in the case of fresh or dried fruit, the number of organisms per food item (Hint: Don't forget the dilution you made).

7. Gram stain at least one interesting colony from each plate.

8. Record the results in **Table 22.1**. Note: include the results for 2 food items in each of the 3 food classes (number 1, 2, and 3 above). It may be necessary to obtain results from other students.

Results

Table 22.1 Colony characteristics and abundance of microorganisms cultured from various types of food. Include the results for 2 food items from each of the 3 food classes tested.

Food Item	Colonies/Plate	Microbes/Gram of Food	# Colony Types	Colony morphology

Gram Stain Results:

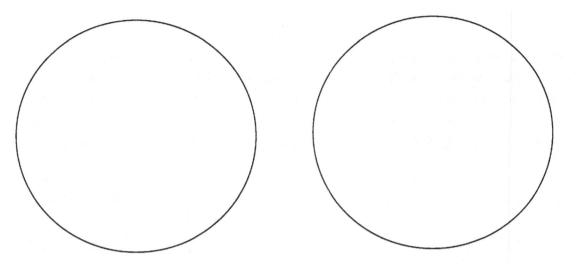

Gram Reaction/Morphology and Food of Origin:

_____   _____

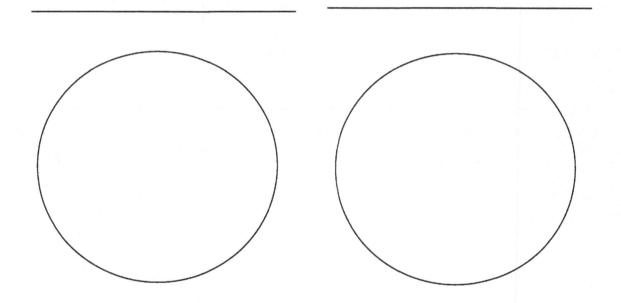

Gram Reaction/Morphology and Food of Origin:

_____   _____

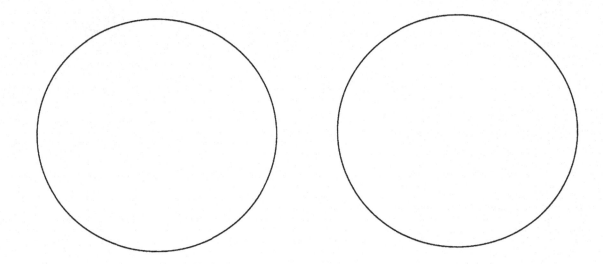

Gram Reaction/Morphology and Food of Origin:

_____   _____

Questions

1. Discuss the significance of contamination of foods that are cooked as opposed to foods that are eaten raw.

2. Explain why it is safe to order your steak rare but not your double cheeseburger.

3. Differentiate between, and give examples, of the two types of food poisoning.

Exercise 23: Milk and Yogurt

Objective: To understand a qualitative method of testing milk and to understand the process of making yogurt. Upon completion of this exercise you should be able to:

- ❖ Define biological oxygen demand (BOD).
- ❖ Be familiar with the principles of pasteurization.
- ❖ Explain the general principles on which standard methods for testing milk are based.

Application: Milk testing, pasteurization and preservation are based on an understanding of microbial metabolism. In the case of yogurt and cheese production, the preservation of milk in these forms is accomplished by a fermentation process carried out by the microbes themselves.

Methylene Blue Reductase Test:
Cows have normal microbial flora that contaminates their milk. These organisms are harmless to calves and to humans with healthy immune systems. However, these bacteria ultimately play a role in the spoilage of milk. Bacteria divide very rapidly and if milk is allowed to stand at room temperature for 4-12 hours, the properties of the milk will change. As fermentation byproducts are released by the metabolizing bacteria, the milk proteins coagulate forming curds and squeezing out water. Since refrigeration slows down bacterial growth and pasteurization reduces bacterial numbers either treatment can be used to increase the time it takes for spoilage to occur.

Pasteurization, a process named after its inventor, Louis Pasteur, relies on heating to reduce the number of organisms present in milk in order to make a safer product. The strategy is to reduce the total number of organisms and thus reduce the number of pathogenic organisms while not significantly changing the character of the product. At the time that pasteurization was developed, cattle were prone to infections of the udder which could be transmitted to humans. Today because we now have a better understanding of microbiology and use antibiotics to treat infections in animals, pasteurization is used more for increasing the shelf-life of the product, than for preventing the spread of disease.

Surveillance methods are used by the milk industry to test their products after pasteurization. Quantitative methods to assess bacterial numbers include plate counts and direct microscopic counts but these tests are laborious. Consequently less laborious qualitative tests are often used. One such test, and the test you will perform, uses methylene blue to detect the amount of oxygen in the milk. Methylene blue is blue only when in the presence of oxygen; in the absence of oxygen, it is colorless. When microbes grow in milk they use oxygen from the milk to carry on aerobic respiration; the removal of oxygen can be followed using methylene blue. The amount of oxygen microorganisms use when they metabolize the organic material in water or milk is called Biological Oxygen Demand (BOD). If methylene blue is added to a test tube containing milk, since the milk is oxygenated, the milk turns blue. The tube is then incubated undisturbed for 8 hours at 35°C. As organisms in the milk grow, they use up the oxygen and the blue color disappears. The length of time it takes is indicative of how many organisms are present in the milk at the start of incubation.

Procedure:

1. Add 10 ml of milk to a large screw cap test tube.
2. Add 1 ml of methylene blue reagent and invert to mix the tube.
3. Place the tubes in a 35°C water bath, incubate 5 minutes and invert to mix. DO NOT MIX AGAIN.
4. Observe the tubes every 30 minutes for the next eight hours.

Record results in the **Table 23.1** in the **Results** section. Include the class results from 3 different milk samples. Results should be classified in the following way:

Class 1: Excellent, not decolorized in 8 hours
Class 2: Good, decolorized in less than 8 hours, but not less than 6 hours
Class 3: Fair, decolorized in less than 6 hours, but not less than 2 hours
Class 4: Poor, decolorized in less than 2 hours*

*most non-pasteurized milk will decolorize in about 2 hours.

Making Yogurt:

In another aspect of the dairy industry, microbes are used to create food products such as yogurts and cheeses. Originally, yogurt and cheese producers relied on the normal flora of the milk to carry out the fermentation. Milk was incubated at a temperature of around 45°C, the optimum growth temperature for the primary organisms that perform the desired fermentation reaction. Because this temperature is so different from the incubation temperature of most pathogens, it is a safe and effective way to make yogurt; this method is still used in many parts of the world.

To increase the efficiency of the fermentation process and to insure safety from pathogens, yogurt manufacturers in the United States begin the process with pasteurized milk. They then inoculate the milk with large numbers of lactose fermenting organisms grown in pure culture. (*Lactobacillus acidophilus* is commonly used but is often mixed with other fermenters). This ensures that the predominant organisms in the yogurt are the fermenters of choice and that the fermentation goes to completion as fast as possible. Making yogurt is a means of preserving milk because the acid produced by the fermentation process lowers the pH of the product to a level that inhibits the growth of spoilage organisms.

1. Weigh out 4 grams of dry powdered milk and dissolve in 100 ml of fresh whole milk.
2. Heat the milk to boiling. Stir constantly to prevent burning and remove from heat as soon as the milk reaches the boiling point.
3. Cool milk to 45°C. (The milk will be cool enough when you can hold the beaker in your hand without burning)
4. Add 1 teaspoon of yogurt. (The yogurt package should indicate that it contains active culture such as live *Lactobacillus acidophilus*).
5. Cover the beaker and incubate at 45°C overnight.
6. Yogurt is now ready to mix and consume.
7. Taste the product before adding sugar, then add sugar and/or fruit and EAT.

Results

Table 23.1 Methylene blue reductase test results.

Milk Sample	Source	Result
1		
2		
3		

Describe the consistency and taste before and after adding sugar of the yogurt that was made in the lab. What do you think accounts for the tart taste of the unsweetened yogurt?

Questions

1. Describe how normal flora present in milk contributes to spoilage.

2. How does refrigeration slow down the spoilage process?

3. Describe the process of pasteurization.

4. Describe how the methylene blue reductase test works when testing milk quality.

5. What is the purpose of heating the milk prior to inoculation with the yogurt culture?

6. What is the purpose of adding a teaspoon of yogurt to the milk prior to incubation?

7. Why is the culture incubated at 45°C?

Exercise 24: Public Health and Drinking Water

Objective: To become aware of the clean water regulations and several methods for routine testing of our water sources. Upon completion of this exercise you should be able to:

- ❖ Explain the function of indicator organisms.
- ❖ Name the organism which is typically used as an indicator of fecal contamination in water.
- ❖ Explain why this organism is used.
- ❖ Explain which components of EMB agar are selective and which are differential.
- ❖ List the maximum number of *E. coli* permitted in 100 ml of drinking water in the US.
- ❖ Explain the filtration method of screening water samples.
- ❖ Explain the Colilert method of screening water samples.

Application: Fecal contamination of water is a world-wide concern. In many developing countries, public health practices are insufficient to maintain clean water and in some countries, cholera and typhoid fever epidemics are common.

Public health practices in developed countries, such as the U.S., include sewage treatment procedures that ensure that our water supplies are free of sewage contamination. Routinely our water supplies are checked for contamination by testing for *E. coli*. While most strains of *E. coli* are nonpathogenic when ingested by humans, *E. coli* is used as an indicator of fecal contamination since it is found in the intestines of all humans and most mammals. Indicator organisms of fecal contamination, such as *E. coli*, are used to trace fecal contamination and, by implication, indicate the possible presence of human pathogens that can be transmitted via the fecal-oral route.

Standard protocol in the U.S. states that 100 ml of drinking water should contain no *E. coli*. Any deviation from the normal is considered unacceptable. However, numerous environmental organisms are acceptable and in some cases desirable for a healthy water supply.

There are several methods for rapidly and inexpensively determining if *E. coli* is present in a water supply. In this laboratory exercise two of these methods will be used to demonstrate one of the functions of our public water utilities. The first screening method relies on a filtering apparatus to collect the bacteria from 100 ml of drinking water. These bacteria are then grown on eosin methylene blue agar (EMB). EMB is both selective and differential. It selects for enteric organisms because it contains bile salts to which most enterics are resistant given their natural habitat, the bowel. It is differential because it contains the sugar lactose. Colonies that can ferment lactose appear pink while those that do not appear colorless or the color of the medium. And finally what makes this media perfect for water testing is the fact that colonies of *E. coli* have a green sheen when grown on EMB.

The Colilert method is commonly used by public health facilities. It is rapid to set up and can easily distinguish *E. coli* from other enterics. The Colilert uses two substrate molecules with colored products to test for the activity of two enzymes. The first of these enzymes is β-galactopyranosidase, an enzyme involved in metabolism of the disaccharide galactose. The substrate used in the Colilert procedure is similar to galactose except p-nitrophenol has replaced one of the sugars of the disaccharide. If the enzyme -galactopyranosidase is present, it will break the bond between the sugar and the p-nitrophenol. Once the bond is broken, the p-nitrophenol gives a bright yellow color. This indicates that lactose fermenters are present. Further, there is another enzyme substrate for the enzyme β-glucoronidase, which is **unique to *E. coli*.** When this substrate is broken by the enzyme, a fluorescent product, methylumbelliferone, is released. Methylumbelliferone can be detected under ultraviolet light.

Procedure

Method 1: Filtration
1. Using the vacuum pump and filtration apparatus, filter 100 ml of water sample.

2. Aseptically remove the filter from the apparatus with sterile forceps.

3. Place the filter on the surface of an EMB plate, taking care not to trap air bubbles under the filter.

4. Incubate the plates at 35-37°C for 18-24 hrs.

5. *E. coli* colonies can be distinguished by the green sheen that appears only with *E. coli* colonies on EMB.

6. Count the colonies and determine the number of organisms per ml of water (Hint: Don't forget the dilution you made).

7. Record the results in **Table 24.1**.

Method 2: Colilert
1. Place 100 ml of water in the Colilert bottle.

2. Add the packet of Colilert medium.

3. Mix well and incubate overnight at 35-37°C.

4. Look for a visible color change to yellow.

5. Observe all yellow bottles under the UV light.

6. Record the results in **Table 24.2**.

Results

Table 24.1 Filtration experiment results.

	Source of water sample	# Colonies/plate	# Bacteria/ml	# Colony Types
Control				
Water Sample 1				
Water Sample 2				
Water Sample 3				

Table 24.2 Colilert experiment results.

	Yellow	Fluorescent
Control		
Water Sample 1		
Water Sample 2		
Water Sample 3		

Questions

1. Generally, what is the function of indicator organisms? Which organism is typically used as an indicator organism in water quality testing and why?

2. Describe the selective properties of EMB agar.

3. Describe the differential properties of EMB agar.

4. According to the US standards for potable water, what is the maximum acceptable number of indicator organisms present in 100ml of drinking water?

Exercise 25: Fungi

Objective: To become familiar with several common fungi and fungal diseases. Upon completion of this exercise you should be able to:

❖ Differentiate between molds and yeasts.
❖ Define superinfection and give an example.
❖ Explain how fungi are identified.
❖ Describe how you would differentiate between a yeast colony and a bacterial colony.
❖ List several locations where fungi are normally found.
❖ Define dimorphic fungi.

Application: Several of the common infectious diseases in the U.S. are caused by fungi, especially yeasts.

Yeasts and molds are eukaryotes which belong to the group of organisms referred to as fungi (fungus, singular). Yeasts occur as single cells while molds are primarily multicellular filamentous organisms. Some fungi, referred to as dimorphic, display both morphologies depending on environmental conditions. Dimorphic fungi growing at room temperature form multicellular filaments, but when growing in the body at 37°C they occur as single-celled yeasts. Yeasts grow in a manner similar to bacteria, producing a creamy colony on agar media. Molds, on the other hand, grow as large filamentous colonies, such as those seen on old bread. There are numerous human pathogens in both groups of fungi.

Candida albicans is a pathogenic yeast. It is often the cause of vaginal infections when the pH of the vagina is raised toward neutrality. It is also the cause of thrush, an infection of the mouth and can cause overgrowth of the colon when antibiotics have killed off the normal bacterial residents; this is called a superinfection.

When *Candida albicans* grows on solid media it is indistinguishable from bacterial colonies until Gram stained. Yeast cells are much larger than bacterial cells. *C. albicans* is distinguished from other pathogenic yeasts by biochemical tests and the formation of germ tubes when incubated in serum for 2-4 hours. (Germ tubes are pseudohyphae or a form of budding reproduction, which are unique to *Candida albicans*)

Molds such as *Aspergillus niger* and *Penicillium notatum* are considered common harmless contaminants of food, especially bread. However, in some cases such as impaired immunity, *Aspergillus niger* can be an opportunistic pathogen infecting the lung and other parts of the body. *Aspergillus niger* is a rapidly growing, hearty mold from the soil that will grow on most types of media. It first appears as a white filamentous growth which rapidly turns black as the organism sporulates. *Penicillium notatum* is famous for its production of the antibiotic penicillin. Several species are used to produce the distinctive flavors of specialty cheeses like camembert. *Penicillium* sp. are ubiquitous in many environments such as soil and the dry spores they produce can be carried in the air. As a consequence these species are common contaminants of food and can contaminate microbiological media. They begin as a white filamentous growth that may not even be noticed on bread or other foods and when mature the green, beige, or blue spores formed become obvious. When grown on agar, the surface of the colony often appears wrinkled.

Identification of fungi is based on colony morphology and microscopic characteristics (e.g. the type and arrangement of sporing structures). *Aspergillus* produces spores in chains on top of bulbous conidiophores. *Penicillium* produces chains of spores in brush-like structures. Slides must be prepared carefully to allow observation of the spores and sporing structures.

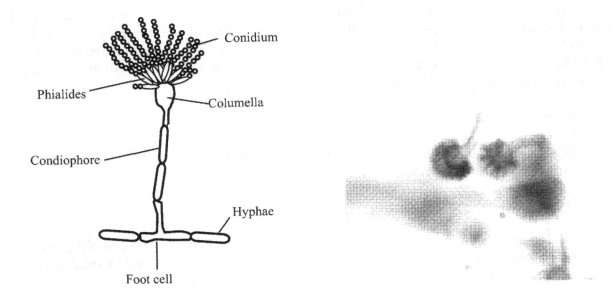

Figure 25.1 Aspergillus

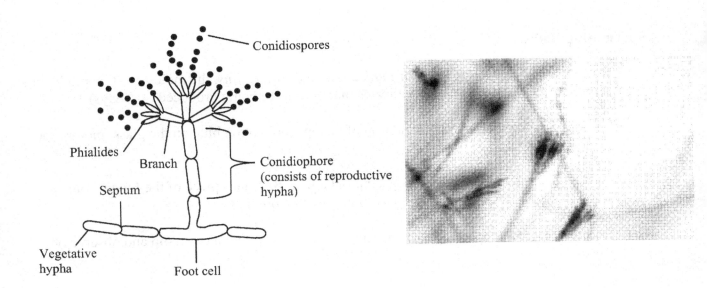

Figure 25.2 Penicillium

NOTE: Because yeasts are single-celled, when Gram stained they can be confused with bacteria. They will appear as large round to oval cells that stain variably (purple and pink on the same cell). However, these cells are very much larger than bacteria. To avoid confusion, note the magnification used to observe them. Under 400X magnification a yeast cell will look as big as your average bacterium does at 1000X.

Germ Tubes

1. For the production of germ tubes, pseudohyphae, inoculate serum (bovine, fetal calf, horse, etc.) with a small amount of fresh *Candida albicans*.

2. Incubate the tube for 2 hours at 37°C.

3. Remove the tube, prepare a wet mount of the resulting culture and observe for germ tube formation using the 40X objective. If none are seen the culture should be incubated another two hours and observed again, before calling it negative.

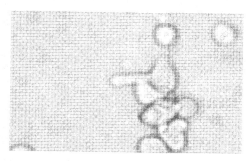

Figure 25.3 *Candida albicans* germ tube

Sporulating Bodies

1. To observe the sporing body of *Penicillium* and *Aspergillus*, the molds must be producing spores. The color formation indicates that spores are being produced (4-7 days).

2. Slides are prepared by putting a drop of lactophenol cotton blue on the glass microscope slide.

3. Using a sterile scalpel and forceps, cut out a tiny (3x3 mm) piece of the fungal colony and place it in the drop of lactophenol cotton blue on the slide.

4. Tease the fungus apart using the teasing needle, cover with a coverslip and observe on 40X.

Results

1. Draw *Candida albicans* germ-tube.

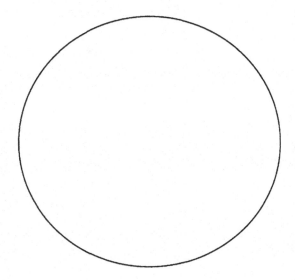

2. Draw *Penicillium* and *Aspergillus* as seen on your lactophenol cotton blue stained slide.

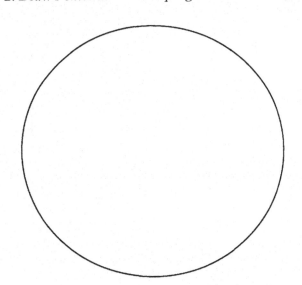

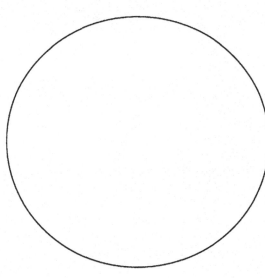

Name _____   Name_____

Questions

1. What two morphologies are found in the group collectively called fungi? Differentiate between the two.

2. What is a dimorphic fungus? Why is it significant to human health?

3. Define superinfection.

4. How are fungi identified?

5. How would you differentiate between a colony of *Candida albicans* and a similar looking colony made up of bacterial cells?

Exercise 26: Antibiotic-Producing Microbes from the Soil

Objective: To understand the origins of antibiotics. Upon completion of this exercise you should be able to:

- ❖ Make dilutions of soil for the purpose of enumeration of microbial colonies.
- ❖ Describe where most antibiotics originate.
- ❖ Describe the natural purpose of antibiotics from soil organisms.
- ❖ Describe microbial diversity.
- ❖ Describe some of the uses of soil microbes.

Application: The majority of the antibiotic-producing organisms come from the soil where competition for nutrients and growth factors is thought to be the major driving force behind their production.

An important component of soil is the abundant and diverse array of microorganisms which it supports. These soil microbes are extremely important to our lives. Nitrogen fixers in soil make atmospheric nitrogen available to plants by converting it to nitrates and nitrites. Decomposers, by degrading dead plant and animal material, recycle biologically important molecules.

Soil is an important source of antibiotic-producing microorganisms. These organisms secrete the antibiotics into the soil where they inhibit other organisms in the immediate vicinity. Studies indicate that most of the antibiotic producing bacteria and fungi are slow growing organisms. Based on this observation, it is thought that the release of antibiotics offers an advantage to the slower growing organism that would otherwise be overgrown by the more rapidly reproducing bacterial species.

Interesting, too, is the question of when and where antibiotic resistance arose. Evidence suggests that the resistance mechanisms that are commonly encountered in medicine originated in soil organisms. The use of antibiotics in medicine has served to select for organisms that have resistance mechanisms. This is a serious problem in therapy for bacterial infections.

In response to the increase in antibiotic resistant strains of disease causing microbes, researchers have returned to the soil to search for new natural antibiotics. The following exercise demonstrates how the soil is mined for new drugs.

Procedure:

Day 1:

1. Obtain two 9.0 ml water blanks and two 9.9 ml sterile water blanks.

2. TUBE 1: Add one gram of soil to be tested to 9.0 ml of sterile water and mix well.

3. TUBE 2: Transfer 0.1 ml from TUBE 1 to a 9.9 ml water blank. Mix thoroughly.

4. TUBE 3: Transfer 0.1 ml from TUBE 2 to a 9.9 ml water blank and mix thoroughly.

5. TUBE 4: Transfer 1.0 ml from TUBE 3 to a 9.0 ml water blank and mix thoroughly.

6. Transfer 0.1 ml from TUBE 3 to the surface of a glycerol yeast extract agar plate and spread with a sterile hockey stick.

7. Transfer 0.1 ml from TUBE 4 to the surface of a glycerol yeast extract agar plate and spread with a sterile hockey stick

8. Allow time for the plates to dry before inverting and incubating at 30°C for 7 days.

9. After incubation, and with the help of your instructor, look for *Actinomyces*-like colonies. These colonies will have a dusty appearance due to the presence of spores.

Day 2:

1. Streak a TSA plate for confluent growth with *Staphylococcus epidermidis* using a swab.

2. Divide the plate into quadrants with a sharpie.

3. In the center of each quadrant, inoculate one of 4 suspected *Actinomyces* colonies. Make the inoculation by just touching the inoculated loop to the surface of the agar. Be sure to number the isolated colonies from the original plates so that the antibiotic producers can be isolated later.

4. Incubate the plates at 30°C for 7 days.

5. Observe the plates for zones of inhibition around the suspected *Actinomyces* colonies.

6. Record the results in **Table 26.1**.

Results

Describe the colonies isolated on the soil plates.

Draw the gram stain results for two of the isolates.

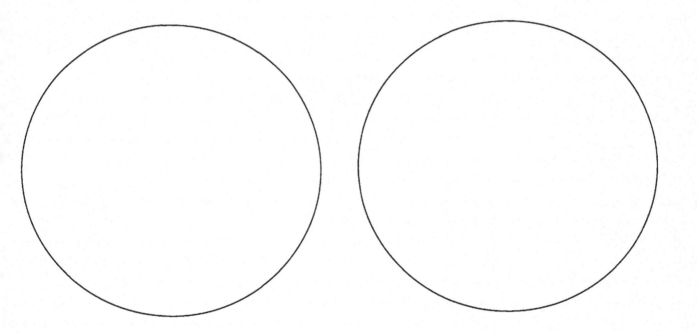

Table 26.1 Record results of antibiotic susceptibility tests performed on Day 2.

Actinomyces Isolate	Zone Size (mm)
1	
2	
3	
4	

Appendix A

Internet Scavenger Hunt

The following exercise is to aid you in becoming familiar with several permanent websites that we feel will be of value to you throughout your career. The purpose is to find the website, lookup the subject of the question, and answer the question. You should take the opportunity to look around the various sites and see what is available for your future use.

1. What are alternatives to penicillin G treatment? (www.cdc.gov)

2. Define the term "emerging pathogen"? (www.cdc.gov)

3. What does MMWR stand for? (www.cdc.gov)

4. Name a chemical and a microbial agent on the Emergency Preparedness and Response list. (www.cdc.gov)

5. What are the immunizations for a four month old baby at a "well-baby visit" to the pediatrician? (www.cdc.gov)

6. Name the two fellows that discovered *Helicobacter pylori*? What disease does this organism cause? (www.cellsalive.com)

7. What is ATCC? (http://www.atcc.org/)